# Lost
# Antarctica

# Lost Antarctica

## ADVENTURES IN A DISAPPEARING LAND

## James McClintock

palgrave
macmillan

First published in 2012 by PALGRAVE MACMILLAN® in the United States—a
division of St. Martin's Press LLC, 175 Fifth Avenue, New York, NY 10010.

Where this book is distributed in the UK, Europe and the rest of the world, this is
by Palgrave Macmillan, a division of Macmillan Publishers Limited, registered in
England, company number 785998, of Houndmills, Basingstoke, Hampshire RG21
6XS.

Palgrave Macmillan is the global academic imprint of the above companies and has
companies and representatives throughout the world.

Palgrave® and Macmillan® are registered trademarks in the United States, the
United Kingdom, Europe and other countries.

ISBN 978-0-230-11245-2


Library of Congress Cataloging-in-Publication Data

McClintock, James B., 1955–
   Lost Antarctica : adventures in a disappearing land / James McClintock.
      p.   cm.
   Includes index.
   ISBN 978-0-230-11245-2
   1. Antarctica—Environmental aspects. 2. Antarctica—Description and travel.
I. Title.
G860.M39133   2012
559.89—dc23

                                                                        2012005626


A catalogue record of the book is available from the British Library.

Design by Letra Libre, Inc.

First edition: September 2012

10  9  8  7  6  5  4  3  2  1

Printed in the United States of America.

*For Ferne, Luke, and Jamie McClintock*

# Contents

*Eight pages of photographs appear between pages 114 and 115.*

# *Acknowledgments*

*T*his book owes its inception to my friend, colleague, and fellow Alabamian, Edward O. Wilson, Pellegrino University Professor at Harvard University. Ed saw the potential and encouraged me as a scientist and naturalist to write for a popular audience. I am indebted to Ike Williams, and especially to Katherine Flynn, my literary agent, for their encouragement and guidance as I developed my book proposal and authored a sample chapter.

I thank my editor at Palgrave Macmillan, Luba Ostashevsky, for her generous patience with a first-time author, and for skillfully and artfully providing editorial guidance. I also extend a special thank you to Adam Vines, poet, author, and assistant professor of English at the University of Alabama at Birmingham, for his unflagging editorial input, his keen sense of the literary, and a shared love of fishing. I should thank a host of others at Palgrave Macmillan, including Carla Benton (assistant production editor), Laura Lancaster (assistant editor), David Rotstein (art designer), Siobhan Paganelli (publicist), and Christine Catarino (marketing manager). Ryan Masteller did a superb job of copyediting. I am grateful to a number of friends and colleagues who reviewed portions or full drafts of my book manuscript: Margaret Amsler, Joseph Shepherd, James Bockheim, John Lawrence, Susan Solomon, Henry Pollack, Hugh

Ducklow, Marilyn Kurata, and Janice O'Reilly. Bill Fraser, Bill Baker, Charles Amsler, Richard Aronson, Sven Thatje, Donna Patterson, and Jen Blum kindly provided answers to technical questions.

My gratitude is extended to those exceptional individuals who read an early draft of my book manuscript and provided a jacket blurb: Edward O. Wilson, Sylvia Earle, Elizabeth Kolbert, Henry Pollack, Susan Solomon, Hugh Ducklow, Douglas Brinkley, and Bill Gates, Jr. Bill Gates, Sr., and John Pinette helped facilitate the jacket blurb by Bill Gates, Jr.

My career as an Antarctic marine biologist would not have been possible without my academic mentors John Lawrence and John Pearse inviting me to carry out Antarctic research and the support provided by the Office of Polar Programs (OPP) at the National Science Foundation (NSF), Washington, D.C. Thanks to OPP Program Officer Polly Penhale for funding my first grant and supporting my research program for many subsequent years. Thanks also to Roberta Marinelli, who subsequently stepped in as OPP program officer and continued the tradition of providing expert guidance. OPP Division Director Scott Borg was also helpful as was Peter West, OPP Polar Education and Outreach Program officer, who continues to facilitate Charles Amsler and my Antarctic educational outreach programs. I also greatly appreciate the employees of the contractors (ITT, Antarctic Support Services, Raytheon Polar Services Company, Lockheed Martin) who provide logistical support for the U.S. Antarctic Program. There are far too many support staff spanning the past three decades to thank by name, but Robert Farrell and Rebecca Shoop, who both serve as current Palmer Station managers, deserve special recognition.

I wish to acknowledge Robert Fischer, chair of the Department of Biology at the University of Alabama at Birmingham (UAB) and

Carol Pierce (office services specialist) for departmental support, as well as fellow Biology faculty members Ken Marion, Robert Angus, and Stephen Watts, who have assisted on the home front while I was in Antarctica. Thanks are due to Dale Turnbough, Andrew Hayenga, Kevin Storr, Deborah Lucas, Linda Gunter, Pam Powell, and Jo Lynn Orr in UAB Media Relations and Becky Gordon and Jennifer Ellison in UAB Development. Shirley Salloway Kahn, UAB vice president for Development, Alumni, and External Relations, has been very supportive, as has UAB President Carol Garrison.

The Robert R. Meyer Foundation, Birmingham, Alabama, has supported my research by making a generous gift to establish an Endowed Professorship in Polar and Marine Biology at UAB which continues to provide much needed support for my Antarctic marine research. Abercrombie & Kent Philanthropy donates funds to my Antarctic research program, and my thanks and appreciation are extended to Jorie Kent (A&K Philanthropy vice chairman), Geoffrey Kent (A&K chairman and chief executive director), Bob Simpson (A&K VP for Business Development), Pamela Lassers (A&K director of Media Relations), and Julia Evanoff (A&K product operations coordinator). Thanks to Jane Hazelrig (travel consultant), Ann Johnson (president), and Borden Burr (chairman and CEO) at All Seasons Travel in Birmingham, Alabama. All Seasons Travel assists in bookings for the A&K Classic Antarctica Climate Change Challenge cruises I lead each December, and has established a fund to support travel to scientific meetings for my graduate students. Elizabeth and David Wyse have kindly provided support from their foundation for my Antarctic research.

Finally, I thank my wife Ferne, for her grace and patience during my fourteen field seasons in Antarctica. My mother Muriel Ganopole, my stepmothers Evie McClintock and Terry Boles, my brothers Peter and

Michael McClintock, and my sister Melissa McClintock have provided encouragement and emotional support. To date, ten members of my family have traveled to Antarctica with me, and as such, understand its allure. I only wish my father Charles McClintock had lived long enough to make the voyage.

# Chapter 1

## Journeys South

*I* *will never forget the austral spring day thirty years ago when I was* aboard the *Marion DuFresne* as she anchored offshore the Crozet Island Archipelago, an otherworldly cluster of French islands. These islands are nestled deep in the southern Indian Ocean, and their seas, shores, and valleys teem with the iconic penguins, seals, and seabirds that are an integral component of quintessential Antarctica. The sky was crystal clear, the breezes were mild, and the seas were calm. As we approached the shore in our small motor boat, I was taken aback by what appeared at first to be groups of miniature dolphins swimming in orchestrated synchrony in the deep blue waters. As we drew closer, I realized that these were not small dolphins at all, but rather schools of penguins cresting above and sliding below the sea's surface in unison, porpoise-like, to increase speed and save energy while temporarily airborne as they darted back and forth between shore and sea. On the beach, what appeared to be a welcoming party of king penguins assembled before us. They stood about waist-high, second to the world's largest penguin, the slightly taller emperor. These regal birds were distinguished by a progressive gradation of light-yellow to deep-gold feathers that graced the upper reaches of their snow-white chests. On either side of their black heads, tear-dropped patches of gold complemented the gold slashes that adorned the lower sides of their beaks, as if their beaks were painted by a final stroke of a brush for good measure.

After debarking, I wound my way along the beach, past nests with females sitting on their eggs and sidestepping the occasional squawking downy-feathered chick. I worked my way up a steep hillside that framed the mouth of the immense valley where the penguins had set up their colony. Several hundred feet above the valley floor, I stopped to take a photograph of our landing spot with its sprinkle of elephant seals sprawling amid the king penguins. Then, turning to face the opposite direction, I snapped the photograph that would later land on the June 1984 cover of *BioScience*. Over fifty thousand king penguins stretched across the river valley floor, disappearing around the distant bend. Their braying chorus and pungent guano overwhelmed my senses and fully complemented the stunning view. Above me, against the verdant hillsides, a smattering of white dots marked the locations of nesting giant wandering albatross, the largest of the world's seabirds.

My enduring fondness for Antarctic marine biology blossomed during that first expedition south. To this day, I don't know exactly what possessed John Lawrence, my professor and doctoral mentor at the time, to ask me in the summer of 1982 if I would join him on an expedition to the Kerguelen Archipelago, a remote cluster of volcanic islands at fifty degrees latitude approximately one thousand miles off the northeastern coast of Antarctica in the southern Indian Ocean. I have always considered these subantarctic islands, although not in Antarctica proper, to be "Antarctic" because they are teeming with iconic Antarctic seabirds, penguins, seals, and whales. We would study the reproduction, nutrition, and ecology of echinoderms, a wondrous group of marine animals comprised of such familiar creatures as starfish, sea urchins, sea cucumbers, and brittle and feather stars. The research expedition would be jointly funded by the National Science Foundation and the French government (*Centre National de la Recherche Scientifique*). As a graduate student, I was thrilled because of its inherent adventure and also because very little was known about the biology of Antarctic echinoderms or any other group of marine invertebrates at that time. We embodied the scientific equivalent of "kids in a candy store."

With John Lawrence and with great anticipation, we prepared to leave for Antarctica. We first stopped in Paris, where we met briefly with Professor Allain Guille, an expert on echinoderms at the *Muséum d'Histoire Naturelle* who had arranged permission from the French Antarctic Program for John and me to work at the French base on Grand Terre Island in the Kerguelen Archipelago. Very few Americans had ever been to the Kerguelen Islands, and we were the first American scientists to stay on the island since the 1874–1875 American, British, and German expeditions that went there to best observe the transit of Venus: an astronomical event occurring once every 240 years during which Venus passes directly between the sun and the earth. Over a century later, we two Americans would travel there to explore an equally wondrous and mysterious frontier: the depths of the sea rather than the solar system.

The next leg of our journey took us across northern Africa with a brief respite to refuel in the tiny country of Djibouti, bordered by Ethiopia to the west and south, Somalia to the southeast, and Eritrea to the north. We next refueled on the tropical Seychelles Archipelago, a cluster of 115 paradisiacal islands about 900 miles east of the African coast. The plane almost skidded off the runway as we landed in the midst of a torrential downpour at the international airport in the capital city of Victoria. The pilot threw the engines into full reverse and slammed on the brakes. I looked out the window as, still skidding, we closed in on the final few meters of runway. After we stopped, I could just make out a sheer precipice that dropped to the crashing sea.

On the tropical mountainous island of Reunion, just to the east of Madagascar and where John and I would meet our ship, we spent several days at the home of Sonya Ribes, an echinoderm researcher and John's good friend. Her husband Nicholas took us for an unforgettable drive on the steep and narrow winding mountain roads. He drove like a maniac, and though I had to maintain a death grip on the seat in front of me, I

was still able to gape at the gorgeous vistas, flowering hibiscus, bougain-
villea, orchids, bamboo, ferns and palms, and jet-black volcanic sand
beaches scalded by the afternoon sun. Because I had months of cold
weather, endless seascapes, and volcanic islands with limited plant life
ahead of me, I soaked up as much of the warmth and the lush tropical
vegetation as I could take.

John and I spent the next two weeks on the open sea aboard the
*Marion DuFresne*, which was manned by a crew of French sailors and
was bound for the Kerguelen Archipelago, some 1,800 nautical miles to
our southeast. The *Marion DuFresne* is named for an eighteenth-century
French minister and was the first ship built to specifically service the
*Terres Australes et Antarctiques Françaises* (translated from French meaning "Southern and Antarctic Lands"). At 360 feet in length and with a
sleek, deep hull and powerful engines, the beam of the *Marion DuFresne*
cut efficiently and smoothly through the seas. In French society, scientists rank high in stature, and as such the captain invited us to dine with
him and his officers each evening in a special stateroom. These meals
were grand affairs, as the captain and his officers dressed in their finest
formal white uniforms, waiters served us from silver platters, and John
and I were treated to carefully orchestrated selections of fine wines and
aperitifs. Although they were enjoyable at first, these three-hour dinners
soon became interminable. My French was so rusty that I quickly lost
my bearings in the rapid banter, which, alcohol-greased, was too quick
to comprehend. John Lawrence finally came up with the appropriate excuse that lunch was so sufficient in scope that we did not need another
large meal, and we spent the rest of the voyage eating our dinners in the
crew mess.

Before arriving at Kerguelen, the captain treated us to a day-long
visit to the Crozet Islands, an archipelago comprised of six small volcanic
outcroppings. These are uninhabited except for a small research station

called Alfred Faure, manned by about a dozen scientists and support personnel and located on the eastern side of Île de la Possession. While there, the ship's crew resupplied the station with food and fuel using the ship's helicopter to sling loads back and forth in a huge net basket. Later, while the ship rested at anchor in Île de la Possession's Alfred Bay, we went ashore to visit one of the world's largest king penguin rookeries, which is where I truly had my first encounter with Antarctica.

Following our wondrous day at Alfred Bay, our five-day voyage to the Kerguelen Island Archipelago was colored by a growing anticipation of at long last arriving at our destination. Similar to the Crozet Islands, Grand Terre Island in the Kerguelen Archipelago is covered with vegetation, primarily a low-lying plant with dull reddish flowers known as *Acaena magellanica*. Whalers and sealers that visited the island in the nineteenth century released rabbits to ensure they would have a food source if they were ever stranded. Unfortunately, the furry herbivores drove many of the native plants to extinction, but *Acaena*, distasteful to rabbits because of a chemical defense, survived and eventually took over the landscape. The French introduced cats to the island in the 1950s to control the rabbits. This was a monumental mistake. The cats soon learned it was easier to eat the chicks and eggs of birds than to chase rabbits. As I hiked across the Kerguelen countryside taking in views of snow-covered Grand-Ross Peak, which at 6,040 feet is the highest mountain on Grand Terre Island, I watched some of these feral cats and rabbits darting about in the vegetation. Kerguelen, like most islands around the world, suffers from the ecological wrath of introduced species. I would learn later that climate changes are lowering natural barriers to introduced species. For instance, as temperatures rapidly rise, subpolar archipelagos such as Kerguelen are subject to invasions of plants, insects, and birds originating from warmer climates.

John and my home away from home, the station of Port-aux-Français, rested on the edge of the Bay of Morbihan on Grand Terre Island, and its small cluster of lime-green buildings housed about eighty French scientists and support staff. From the small marine biology laboratory perched above the sea, I could look out over a dark blue expanse dense with kelp forests. Late at night, I lay in my bed listening to the dorm windows rattle incessantly as the winds howled near hurricane force. Everyone who visits Kerguelen must make his or her peace with the wind. Thousands of miles of Indian Ocean surrounded us, and, uninterrupted by land, the powerful gales gained force before slamming into the island. Each morning, John Lawrence and I woke early for a breakfast of coffee and fresh baked rolls before striking out by foot across the hills to our study site on the rocky intertidal coast, sometimes leaning into winds so violent that they nearly knocked us backward.

We followed a well-worn path, now christened the *"Promenade des Amerlocks"* (American Promenade). But the official naming of our path apparently has a twist for the literal French translation of *Amerlocks*, which is a joke. The suffix is purposely misspelled as an American might spell the word if he did not know much French. Perhaps the French wondered what these two crazy Americans were doing traversing this path day in and day out despite the inclement weather.

Timing our visit to coincide with low tide, we arrived to find generously exposed boulders, rocky platforms, and tide pools, the latter of which are home to beds of mussels, an assortment of green sea anemones, brown snails with shells resembling pointed top hats known as limpets; chitons, which are flattened, cradle-shaped mollusks, and small, delicate pink sea cucumbers. In the lowest reaches of the intertidal zone, closest to the open sea, we discovered a dense coverage of holdfasts of the massive bull kelp *Durvillaea antarctica*, an organism whose unique honeycombed architecture renders its golden-brown leathery blades tough and buoyant. The energetic surf whipped the twelve-foot long fronds back and forth across the rocks. Remarkably, scientists had never documented this unique marine community. John and I eagerly extended rope transects across strategic stretches of the rocky intertidal zone, marking the area we planned to evaluate, and over the following weeks meticulously measured and described its inhabitants for a scientific paper.

I noticed dime-sized smooth spots or "scars" on some of the boulders where limpets rested at low tide. Running my finger across the rock scars, I was reminded of the smooth texture of human skin over which scar tissue forms after a wound has healed. I wondered if these scars could be evidence of what is referred to as "homing behavior." Some tropical and temperate limpets are known to make rock scars by repeatedly scraping the rocks' textured surface with their conveyor belts of grinding teeth, known as "radulae." After completing their foraging excursions at high tide, they return "home" to their own personal rock scars with each receding tide. This behavior, which may seem like mere decoration, protects them from predators as the smooth rock scars provide a more secure gripping surface for the limpets' muscular rubbery feet. A predatory starfish hoping for a tasty meal would find it difficult to pry a limpet off its scar once it had suctioned itself with its plunger-like grip.

We tested our homing theory by marking the Antarctic limpets with an ink pen and then mapping their daily movements. We concluded, however, that despite their scars they do not home. One reason could be that fewer direct predators of the limpet live in the subantarctic. While scuba diving one day near the station, I discovered limpets and starfish sharing the same fronds of kelp. My follow-up observations of this predator and its prey in the laboratory indicated that when the tube-feet of the starfish touch the limpets, the snails wildly gyrate their cone-shaped shells, like a dog shaking off water, thus dislodging the starfish before rapidly turning and fleeing.

The predator-avoidance tactics of an organism so small might not seem to be important when discussing a warming climate. However, Antarctic limpets have turned out to be a crucial component of the Antarctic web of life, providing nutrients and energy for higher trophic level organisms such as fish and especially seabirds. Should climate warming or ocean acidification cause limpet populations to decline, the consequences for consumers further up the food chain could be severe. Similar to many locations around Antarctica, Grand Terre Island features mounds of limpet shells, or "shell middens," along the shores of Kerguelen, indicating that seabirds such as giant petrels and kelp gulls have consumed vast quantities of snails. I often observed "shell dropping" behavior as I sheltered myself from the high winds against the base of the cliffs bordering the rocky intertidal, as kelp gulls rose high into the air to drop mussels and limpets onto the rock platform. Diving to retrieve their cracked prey, the birds would land on the nearby shell middens to pick the soft flesh from the broken shells.

Antarctic limpets have turned out to be important beyond their status as a crucial component in food webs. My current research on the impacts of ocean acidification on Antarctic marine organisms is exploiting limpets as a model for studying what has widely become known as "the other $CO_2$ problem." The original $CO_2$ problem, the warming of the earth due to the accumulation of $CO_2$ (a potent greenhouse gas), has received considerable attention. The other $CO_2$ problem, the absorption of $CO_2$ by the world's oceans, rendering them increasingly acidic, is less recognized to date, but just as important. As such, just as mice have yielded important insights into understanding human cancers, limpets may play a key role in understanding of the impacts of human-induced ocean acidification.

John and I were overwhelmed with the possibilities for further discoveries that presented themselves in this amazing place. Before our ship departed Kerguelen, we had completed yet another of our scientific objectives, a study of the reproductive habits of an Antarctic starfish that, like a chicken, roosts on its large yolky eggs to protect them as they develop. This type of reproduction, known as "brooding," is an unusual strategy. Unlike their Antarctic cousins, warmer-water starfish release millions of eggs and sperm directly into the sea. Their embryos develop into tiny swimming larvae covered with little hairs or "cilia" that beat in a coordinated fashion to assist in swimming and capturing food particles. Eventually, the larvae settle to the seafloor, and, through a metamorphosis as dramatic as a caterpillar changing to a butterfly, they transform into baby starfish. So why instead do most Antarctic starfish brood their young? The answer may reside in the extraordinary length of time it takes for the starfish to develop at such low temperatures. In warmer seas, marine invertebrate embryos and larvae develop in a few days or, at most, several weeks. In Antarctica, development can take four to six months. Maybe allowing one's offspring to swim about in the Antarctic plankton for months on end rather than under mom's protective arms is too risky. As we sailed north and I ruminated at great length upon what I had observed, it hit me: Kerguelen had worked its spell on me, and I became obsessed with solving the riddles of Antarctic marine life.

*In May 2004, with winter* fast approaching in the Southern Hemisphere, I awoke, midair, after being flung from my bunk in the chief scientist's suite while crossing the notorious Drake Passage aboard the USS *Laurence M. Gould*, a 230-foot research vessel. I collided against the wall with a resounding thud and fell to the floor. A week earlier, I was fairly confident that we'd have smooth sailing on my return voyage from a two-month stint at Palmer Station, the American research base perched on the southern tip of Anvers Island just off the west coast of the central

Antarctic Peninsula, and home to about forty-four scientists and support staff. The station consists of two large, deep-blue buildings, one housing laboratory, an aquarium room, a kitchen and dining area, and some dorm rooms; the other contains a generator facility, store, gym, bar, movie room, and more dorm rooms. Other smaller support and equipment buildings, as well as a pier with a dock, round out the station. I'd lived and worked twelve field seasons of my thirty-year career as an Antarctic marine biologist, and I knew that both the weather and climate in this region are subject to rapid change week by week.

Chuck Kimball, the satellite communications technician at Palmer Station, reassured me that the weather map of the Drake Passage looked promising over the next several days with a high barometric pressure system. Chuck was, among other things, a ham radio aficionado with contacts across the Antarctic continent and regions north. He had spent countless field seasons at Palmer Station, and I took comfort in his words. After all, he was an old-timer and knew what was what. His voice crackled over my handheld radio, warning us that wind speeds were building toward the upper-twenty-five-knot limit for small boats. Next to our station's generator operator, whose expertise ensured we all had heat and electricity, and our two chefs, Chuck's job of overseeing communications was one of the most critical positions at the station. Surely, I joked, he could predict the weather.

At the last minute, the *Gould* had been unexpectedly diverted to Rothera Station, a British Antarctic base a day's voyage south of Palmer Station, to assist in an emergency pick up: two sets of aircraft skis the German Polar Research Program had apparently forgotten and which German scientists desperately needed to salvage their Arctic field research season. As we finally resumed our voyage, our captain received orders from the U.S. National Science Foundation to divert yet again to Vernadsky, a Ukrainian Station located only twenty miles south of Palmer.

This time, we intended to carry out a humanitarian medical evacuation, or "medevac," of a Ukrainian scientist suffering from a chronic bleeding ulcer. Shortly after dark, amid sloppy seas, we plucked our patient, the station medical doctor, and a translator from a small rubber boat, all of whom had been sent out by the station commander to intercept our ship en route. As soon as the three Ukrainians boarded, we handed the small boat's pilot several boxes of fresh fruit and set sail. Unfortunately, by now the weather had grown rough, with mounting seas and a thirty-knot wind. More bad news—the barometer was rapidly falling as we headed into the Drake Passage. We were sailing into one of the infamous regional storms.

The Drake Passage, also referred to by Latin American and Spanish historians as *Mar de Hoces*, is a five-hundred-mile swath of open ocean that connects the southern tip of South America to the South Shetland Islands of the Antarctic Peninsula. It is the narrowest crossing between Antarctica and terra firma to the north. Coincidently, two captains' names compete to grace the title of this body of water, and both captains owe their candidacy to the notoriously wicked southerly gales. In January 1526, the Spanish sailor Francisco de Hoces was at the helm of the eighty-ton *San Lesmes* when she was caught in southerly gale force winds at the mouth of the Strait of Magellan and literally blown south into uncharted waters to fifty-six degrees latitude south. When the *San Lesmes* regained the tip of South America, the ship and all its crew members were lost in a ferocious gale. Almost exactly a half century later, Sir Francis Drake also experienced the wrath of gale-force winds upon entering the Pacific Ocean at the mouth of the Strait of Magellan. Drake's ship and crew had been blown so far to the south that they had inadvertently confirmed the existence of a large body of water below South America. Today, perhaps because Drake garnered great honor and fame after his successful 1577–1580 circumnavigation of the globe, his name is used more broadly to grace this tempestuous body of water.

◦◦

*The storm I encountered* while traversing the Drake Passage was not the first time I had been introduced to heavy seas in the Southern Ocean. During my 1982 voyage to Kerguelen, our two-week passage was interrupted by a ferocious storm that registered eleven on the Beaufort Scale, a long-established metric of conditions at sea. It rather nonchalantly lists weather at level eleven as: 1) *up to sixty-five-knot storm winds*, 2) *very high waves of up to forty-five feet*, and—not surprisingly—3) a *white sea with blowing foam*. Such conditions quickly turned the southern Indian Ocean into a gut-wrenching roller-coaster ride. Even some thirty years later, I remember standing on the ship's bridge, tightly gripping a hand railing and watching the massive, cresting swells with reverent awe as they reached heights equivalent to that of the bridge, some forty feet above sea level. I surrendered myself to my bunk for the next thirty-six hours.

Despite my encounter with that memorable storm, I was unprepared for the catapult from my bunk aboard the *Gould*, an unforgettable experience that slammed me to the floor and left me dazed and cursing my predicament under my breath. I spent that entire night sliding back and forth across the length of my cabin floor, listening to the incessant banging and clattering of doors and furniture coming unhinged. On top of this, I felt our ship tilting a bit further with each wave: first to port, then to starboard. With each seemingly deeper roll, I feared our ship would lean further into the belly of the sea, dipping its main deck below sea level, or worse, turning turtle, sending us scuttling for our bright red Mustang Survival Suits and the lifeboats.

◦◦

*Cold Water Neoprene Immersion Suits*—or Gumby Suits, as I like to call them—are manufactured by the aptly named Mustang Survival

Corporation. Each of us aboard the *Gould* was provided with our own Gumby Suit when we set sail. Before tucking my suit into the space between the top of my clothing cabinet and my cabin ceiling, at the request of the ship's marine projects coordinator (MPC), I joined everyone in the ship's lounge for the traditional cold-water-immersion-suit practice session. On the Mustang Survival Corporation website,[1] a long list of seemingly attractive features presumably helps market the suits, including their "five-fingered gloves for warmth and dexterity," "inflatable head pillow [to support] the neck and provide . . . an optimal flotation angle," "water-tight face seal," "reflective tape on shoulders and arms [to aid] in nighttime detection," their ease "to don in stormy conditions" and my favorite, the "5-mm fire retardant neoprene [that] provides flotation and hypothermia protection." Despite these reassuring features, everyone assembled in the ship's lounge knew full well that should they find themselves wearing said suit in the turbulent waters of the Drake Passage, it would be time to make peace with the world. Water temperatures are just a few degrees above freezing, and even while wearing the survival suit, one succumbs to fatal hypothermia within an hour or two, a process beginning with numb fingers and toes and progressing through shivering, convulsions, mental disorientation, and finally, major organ failure.

In the training sessions, we pulled our bright red suits out of their storage bags and unfurled them on the floor. With suits unzipped, we sat on the floor and slid into the neoprene leggings, each foot finding its home within a floppy non-slip bootie. Now, awkwardly standing, we inched the suit up our bodies, forcing each arm into the appropriate socket, filling each fingered glove with a hand. Sweating, huffing, and puffing, we forced our heads through the neck opening into the rubbery hood over which the water-tight face seal and face shield would ensure our protection from the ice-cold spray of the Drake Passage. Once satisfied that we could locate the light switch on the lamp attached to the shoulder of each suit and—last but not least—find the attached whistle that would signal any nearby floating neighbors or crew members of a rescue boat, we peeled off our Gumby Suits, rolled them tightly, and slid them back in their storage bags. This was the last time any of us hoped to see the Mustang Survival Suit until the next training session on our next voyage across the Drake Passage.

Following our Gumby Suit session, we visited the two lifeboats. Each lifeboat looks like an oblong orange capsule comprised of two bathtubs glued top to top with a little turret sticking out of the top. The boats are situated on small platforms on the upper port and starboard decks four stories above the water line, at the same height as the ship's bridge. Once inside the capsules, occupants are completely sealed off from the ocean, safe and secure within. However, they would have to rely on the small diesel-powered engine, their only method of propulsion, rendering them essentially flotsam at the mercy of the swells. Despite the tight fit, ample room and a sufficient quantity of shoulder seatbelts exist in each lifeboat for everyone aboard. Even so, if the ship is listing hard to one side when the "abandon ship" signal is given, only the boat lowest to the sea, and presumably suspended directly over the water, can be safely lowered by the crew. In an emergency, the crew can quickly drop the lifeboats to the sea in a full free-fall.

After climbing the stairs to the top of the small platform that houses one of the lifeboats, I had to duck my head to avoid banging it into the top of the small hatched door of the lifeboat. Stepping down through the opening and onto one of the passenger benches, I peered into the dimly lit interior. It smelled like plastic and diesel fuel. Several long benches along each side of the boat faced one another, and I crawled down the length of one to find a spot to try on a shoulder harness. The harnesses reminded me of those a NASCAR driver might don for the Daytona 500. A pilot's station situated slightly toward the aft of the lifeboat contained a single chair below a glass windshield, a small steering wheel, levers and buttons to operate the small inboard diesel engine, some radio equipment, emergency medical supplies, packaged food, and drinking water.

The MPC leaned his head in and said that the first thing he gives everyone in an "abandon ship" scenario is a good dose of Dramamine for seasickness. I can only imagine what it would be like to be packed in there like sardines. In heavy seas, the lifeboat gives new meaning to the nickname "vomit comet" ascribed to NASA's zero-gravity astronaut training flights, except in this case it is the nautical version.

*Morning light after a long night* on my cabin floor aboard the *Gould* brought little reprieve from the heavy seas; instead, it revealed a scene of chronic motion and chaos. The adjoining office within my chief scientist suite was a jumble of overturned chairs, and the floor was strewn with papers, books, pens, and coffee cups that had launched from the desk and were either wedged against something or still rolling and sliding about. Later that morning, I managed to get out of my bunk (after my mid-sleep catapulting adventure, I now favored the lower one), gingerly gain my balance, and teeter my way slowly across the cabin and out the door to peek at the lounge across the hall. It was empty, but like my office, it too was a jumble of overturned spinning chairs and de-shelved books and DVDs. Even more remarkable was the ship's galley, located one level down. Condiments had become missiles: ketchup and mustard had splattered on walls, and salt and pepper shakers had spun across the floor like bowling pins. A homemade sign swayed on the galley door: "Closed Until Further Notice." Several days later, our captain said the ship had rolled a gut-wrenching thirty degrees to starboard and thirty degrees to port during the height of the storm, despite his best efforts to handle the rolling seas. During the storm, the captain seemed unshaken; however, he later acknowledged that these were the largest waves he had ever negotiated in the Drake Passage, and yet this confession made me feel somewhat better.

The *Laurence M. Gould*—also known among scientists, ship technicians, and station personnel as the "LMG" or the "Big Orange Tub"—rolls heavily because of its unique design. Named after polar explorer, geologist, teacher, and college president Laurence McKinley Gould (who served as second in command for Adm. Richard Byrd's first Antarctic expedition in 1929–1930), the 250-foot-long ship features an ice-strengthened hull, a beam width of 56 feet, and serves both as a research platform for up to 22 scientists and as the primary resupply and transport vessel for personnel moving between Punta Arenas, Chile, and Palmer Station. Edison Chouest Offshore, Inc. built the ship in Galliano, Louisiana, and the company launched it in 1997. Edison Chouest has an excellent shipbuilding reputation, and the two ships they leased to the National Science Foundation (NSF) to support its polar research operations, the *Gould* and the *Nathaniel B. Palmer*, are but two of hundreds of ships they successfully build and operate. However, when Edison Chouest initially slid the *Gould* into the water, it listed significantly to one side. Concrete was subsequently added to the ship's lighter side for balance, but the engineers feared the ship's hull would ride too deep in the water to provide the stability necessary to dock safely. To help, engineers designed and installed rectangular pontoons to each side of the ship's hull, a unique design among research ships. What the ship gains in stability it loses in quickness, as the pontoons lower the cruising speed and reduce fuel efficiency. More to the point for those of us that sail aboard the *Gould*, this redesign makes the ship susceptible to rolling in swells.

The Drake Passage, the gateway to the Antarctic Peninsula, is quintessentially fickle. Sometimes it's as flat as a Swedish pancake, and those who ply its waters refer to it affectionately as "Lake Drake." These rare crossings generate celebration. The water can look like an immense mirror, reflecting the deep blues of the sky or the white and gray kaleidoscope of clouds. The smooth sea surface reveals life rarely seen in wind or roiling waters. In the northern Drake, basketball-sized jellyfish, silvery flying fish, and rafts of brown kelp—*Macrocystis pyrifera*, floating on pneumatocyst balloons—can be seen, the latter dislodged by violent storms.

A small step-up platform sits on the tip of the bow of the *Gould*. I like to perch there in calm seas, leaning forward into the gentle breezes, inhaling the chilled air, caught up in the timelessness of the moment. Below me, the bow slices through the slate-gray waters, generating gurgling sounds of churned water. The hull slaps and sprays seawater starboard and port, and from the water thrown skyward come first large heavy droplets, then smaller ones, and finally a silent spray of fine moisture that blends with air to form a salty mist. On occasion from that perch, I have witnessed pods, or social or family groups comprised of up to twelve interrelated individuals, of hourglass dolphins. Measuring only five feet from nose to tail, their black and white coloration resembles that of killer whales, albeit in miniature. The name derives from the two "hourglass" patches coating the beak, eyes, torso, and flippers. Like most dolphins, those below me ride the bow wakes, exploiting the pressure wave generated by the ship's bow cutting through the water to cop an effortless ride. Like playful teenagers on surfboards, they roll and twist, alternating sides of the bow, periodically catching air with powerful tail flips.

To encounter a pod of hourglass dolphins is a rare treat for humans. The mammals live only in the Southern Hemisphere, populating the waters of the Drake Passage from southern Patagonia to the South Shetland Islands. Early explorers and whalers referred to these dolphins as "sea cows" as their black and white coloration reminded them of dairy cattle from farms back home. They feed on fish and squid. Over the years, I have probably seen 25 hourglass dolphins, a large number when one considers they have a total population in the Southern Hemisphere of only about 140,000 individuals. This number is miniscule compared with marine mammals such as Antarctic fur seals or crabeater seals that number in the millions. Despite the small size of the natural population, their remote distribution protects them from fisheries. Only a few hourglass dolphins have been reported killed in fishing nets. Nonetheless, scientists who study marine mammals are concerned that human-induced climate change could have profound impacts on their comparatively small, and thus vulnerable, population. One hopes that the lovely hourglass dolphin will not be lost to climate-induced changes.

*I have undertaken fourteen research expeditions* to Antarctica, and it's always a formidable task to organize a team of marine biologists. But before a marine biologist can even consider visiting such a remote continent, one must go about securing the necessary funding and logistical support. The British Antarctic Survey, the French National Antarctic Program, and the Australian Antarctic Division are respective national organizations that offer support. As a U.S. citizen, I turn to the National Science Foundation's Office of Polar Programs. The process of requesting and securing federal funds and logistical support for my collaborative research program on Antarctic marine chemical ecology—and more recently, studies of the impacts of a rapidly changing climate on marine plants, animals, and communities of the Antarctic Peninsula—is laborious, yet democratic. The first and arguably most critical step in this process is the selection of collaborators.

This holds especially true for polar expeditions, where one conducts joint collaborative research. The team literally eats, drinks, sneezes, and sometimes even sleeps, together. One learns his or her colleagues' favorite sports teams, rock bands, political spins, dreams, aspirations, and even deepest secrets. However, one also encounters intimate things that one really wouldn't want to know about colleagues, including body odor, snoring patterns, and prejudices. Fortunately, I have assembled teams over the years whose traits, good and bad, have been tolerable at the very least.

Dr. Bill "Billy" Baker is a marine natural products chemist trained at the University of Hawaii, with whom I teamed in 1989. A connoisseur of Birkenstock sandals and whose hair has never met a comb, Billy's roots were Californian through and through. Now a professor at the University of South Florida, he is the rare chemist who is equally as comfortable in a dive suit as a white lab coat. I remember nervously meeting him for the first time in a restaurant at Los Angeles International Airport. We were on our way to McMurdo Station, Antarctica, the relative megatropolis of Antarctic research stations. Situated on the southern tip of Ross Island in McMurdo Sound, up to a thousand Americans populate the eighty or so buildings whose hodgepodge layout resembles that of a small mining town. The buildings range in size from the tiny ham radio shack to three-story structures that house dormitories.

To my relief, Billy and I had great chemistry. Over a beer or two, we quickly struck up a friendship that has turned into two decades of professional collaboration. Had the foundation of our relationship been anything but solid, we could have lost a field season to interpersonal issues. Relationships can be fragile as scientists work long hours on the ice. One year at McMurdo Station, I observed three well-established marine researchers kick off with great aplomb the first of two jointly funded

Antarctic field seasons. When they returned for their second field season they hardly spoke a word to one another. The stress of isolation that scientists and support personnel are subject to as they work in remote polar environments clearly divided this team. Everyone deals with this stress in a unique way: some maintain a rigorous schedule of work, exercise, and sleep; others take time off to be alone; and still others socialize, write, or call family and friends at home. Tempers can grow short, and in some instances people can become violent. A cook once became so agitated with his kitchen assistant that he attacked him with a hammer. Fortunately, the assistant sustained only minor injuries, but the cook was sequestered in his room at the station for the balance of the winter. He was deported under FBI guard to Hawaii as soon as spring flights from McMurdo Station had resumed, and there he stood trial and served a three-year prison term.

Dr. Charles "Chuck" Amsler and I have worked together for many years as well. I was familiar with Chuck's research on the kelp forests of California and the algal communities of Antarctica when I first met him while he was interviewing for a faculty position in my department. Chuck has the build of an athlete who craves thirty-mile bike rides, and he is trained in the field of phycology, the study of the biology of seaweeds. With a doctorate from the University of California (in my hometown of Santa Barbara), Chuck has a superb grounding in marine biological research and considerable experience as a research diver. His knowledge of marine plants and his keen interest in marine chemical ecology, combined with several research expeditions to the Antarctic Peninsula, made Chuck a natural fit for our expanding Antarctic Marine Chemical Ecology Research Program. Even better, Chuck and his wife Maggie, who is also an established Antarctic marine researcher who later became my research associate, both joined the Department of Biology at the University of Alabama at Birmingham where I was an associate professor in 1995. Years later, Amsler Island, located just northwest of the U.S. Palmer Station on the Antarctic Peninsula, would be named in their honor.

Once a researcher's collaborators have been identified, the next step in developing a successful research proposal in Antarctic marine biology is to address cutting-edge topics that can only be addressed in Antarctica. Few locations on the planet are more complicated and costly when it comes to scientific logistics, and as such, research topics must be field specific. Because of challenging weather, unpredictable sea ice or icebergs, and rigorous boating and scuba conditions, the proposed field work must be laid out in meticulous detail. Questions in the National Science Foundation's Operational Requirements Worksheet (ORW) include such diverse items as "Will you be constructing a field camp requiring full-time personnel?"; "Will you be conducting perturbation experiments that require manipulating the habitat of birds or mammals?"; "Will you be generating any hazardous materials?"; and, a personal favorite, "Will you use explosives?—if yes—please add details in the Description Box." Permits are also required for a long list of activities, including the collection of animal or plant material, viruses, bacteria, cell cultures, rock samples, soil samples, marine sediment samples, ice samples, seawater samples, and, yes, even air samples. Despite highly favorable peer reviews of a research plan and operational requirements, the odds of receiving funding for a given Antarctic grant are poor at best. Success rates range from 15 to 25 percent. Many excellent proposals go unfunded. Fortunately, over the years, my own research projects on marine invertebrate reproduction, chemical ecology, and ocean acidification have been consistently funded by the National Science Foundation. I attribute this success to the work of my outstanding research collaborators, the preparation and knowledge of how best to design experiments in this remote and challenging environment, and the designing of projects around timely questions that can be answered only in Antarctica.

Given these sobering odds of having a grant proposal funded, I was exhilarated in the fall of 2008 when Chuck, Billy, and I received an email from Roberta Marinelli, our polar programs officer. "Good news at this point," read Roberta's message. "I am putting your proposal forward for logistical review." We knew this language was tantamount to Roberta's recommending our grant for full funding for another three years. Having unwittingly read Roberta's email at the same time, Chuck and I erupted with tandem shouts of joy from our adjacent offices. After the requisite high fives, we called Billy in Florida. "We're headed back to the ice," I chimed. Chuck and I bounced upstairs and shared the news with our department chair. Bud Fischer, a jovial ecologist who, as department chair, delighted in the successes of his faculty, vigorously shook our hands and joked about the biology department losing us both to Antarctica once again. In the big picture, Bud was happy for us, especially as a portion of the grant funds would be used to support departmental expenses associated with our Antarctic research. Last but not least, I called my wife, Ferne, with the good news. Her congratulations flowed from the heart but were tempered by the realization that I would once again be spending time away from her and our two children, Luke and Jamie.

Even in our moment of celebration, Chuck, Billy, and I knew we wouldn't be heading to Antarctica any time soon. It would take over a full year to lay the groundwork before we could board a flight to Punta Arenas, Chile, to meet the ship. We first had to complete a long, excruciatingly detailed Support Information Package (SIP), in which we'd provide information regarding such necessities as instruments, boating requirements, operational dive plans, and computer technology, as well as lists of the chemicals and general lab supplies we would need to support our proposed research. We dutifully submitted the completed SIP to the Polar Services branch of Raytheon, the large American corporation that, until recently, held the contract for operational support services for the U.S. Antarctic and Arctic programs. As a testament to the immense scale of the U.S. investment in Antarctic and Arctic research, this contract, now held by Lockheed Martin, is the largest nonmilitary government contract in America.

Six months before we departed, I had to submit to my least favorite aspect of the Antarctic preparation process—the dreaded medical exam. Raytheon Polar Services requires that all participants pass exhaustive physical and dental examinations prior to each deployment. "Exhaustive" is an understatement. Knowing full well that most people only go to the doctor when a loaded gun is held to their temple, I have to believe that as an Antarctic marine biologist I am one of the most well-studied specimens in the Northern Hemisphere.

After my perfunctory thirty minutes in the waiting room, I handed my physician (and now great friend) Dr. Joe Shepherd the Raytheon Polar Services form letter warning that Antarctica is the highest, driest, and coldest continent on earth, that emergency situations stipulating hospitalization or complex diagnostic procedures require evacuation to South America or New Zealand, and that at Palmer Station, weather conditions can delay getting a plane in and out, even in the summer. Winter evacuations are impossible, the letter reads. Treating chest pain, acute abdominal pain, or renal calculi (a malady I was unfamiliar with until I discovered it simply means kidney stones) can be a major dilemma.

Dr. Shepherd probably felt some modicum of pity for me. Sitting on the examination table, my blood pressure climbing, I discussed with Joe the long list of "routine tests" that we had to schedule. He fired off a series of questions—"Do you have fibromyalgia? Thyroid Disease? Gout? Angina? Supracentricular Tachycardia? Atrial Fibrilation? Deep Vein Thrombosis? Herpes or Syphilis?" None of these conditions would help me gain medical clearance for Antarctica. More questions: "Have you ever been treated for Schizophrenia? Depression? Bipolar Disorder? Panic Attacks? Anxiety Attacks? Obsessive Compulsive Disorder or Suicide Attempts?" These are bad news as well. If one plans to spend the winter in Antarctica, a psychiatrist must perform a much more elaborate psychological examination.

After enduring the procedures of a routine physical examination, I underwent a chest X-ray, then was hooked up to a bundle of electrodes for an electrocardiogram. A urine sample followed, then a medical technician extracted blood for a variety of tests: analysis of my triglycerides, alkaline phosphatase, potassium, bilirubin, calcium, chloride, cholesterol, creatinine, glucose, aspartate transaminase, sodium, uric acid, iron, hepatitis B and C, and for good measure, my PSA. This last test was mandatory, as my father had died of prostate cancer. Sadly, he passed away before I had the chance to take him to Antarctica. The HIV test was recommended but mandatory only for those who winter over in Antarctica. I did one anyway; it would allow me to donate blood while in Antarctica should someone need it in an emergency. Being over the age of fifty, I had to pass a stress test each time I deployed to Antarctica, a lengthy procedure that involved rapidly walking on a treadmill and then having a radio-labeled compound injected in to my blood. This facilitated an evaluation of my blood flow as it is moved through my heart. I recalled from past years that this involved a rather ominous imaging machine that I lay beneath as still as a cadaver.

Compared to the physical exam, dental exams prior to deployments to Antarctica are relatively benign. They can be foreboding in their own quiet way, though. I recently discovered that the real reason for full-bite wing X-rays each deployment has little to do with the health or treatment of teeth during a stay in Antarctica. Turns out they provide a foolproof means of identifying a patient's corpse should identification be otherwise impossible. In a similar vein, I used to be issued a metal chain with a dog tag bearing my name before flights to McMurdo Station from Christchurch, New Zealand, in the event that our plane went down during the journey. A separate set of toe tags rattled in my backpack. In recent years, the U.S. Antarctic Program has ceased this practice, probably due to advances in the use of dental imaging to identify accident victims.

I usually find myself juggling the emotional and practical ramifications of leaving my family about one month before a departure to Antarctica. Despite having done so fourteen times over the past thirty years, it never gets any easier. Some scientists cope with long field seasons by adopting "ice wives" or "ice husbands" while they are away from their spouses; others who are single seek out relationships only to see them fall apart almost as soon as they head home. Some people last only one field season, and a small number depart almost immediately after discovering that the isolation is not for them. Arranging for someone to mow the lawn and to be there in a pinch for Ferne, Luke, and Jamie is the easy part. Missing Luke's school band performances, Jamie's annual dance recital, and my weekly lunch on the town with Ferne is much harder. In the last few weeks before my departure, family emotions tend to get a bit more testy than usual around the house, but it's also a time that brings us closer together. We already miss one another. Ferne and I never sleep well the night before my flight. On the morning of my departure, hugs for Luke and Jamie are uncomfortably tight before they head off to school. Ferne will drive me to the Birmingham airport, and after a long hug and kiss in the drop-off zone, I pull away and see her tears. Mine follow. "I love you," we'll say to one another. Even before she drives off and disappears around the corner, I start counting the days, weeks, and months until we are reunited.

❧

*Twenty-eight years ago,* while I was a graduate student living and working at the Port-aux-Français station on the Kerguelen Archipelago, my
only means of communicating with Ferne, my fiancée at the time, was
by Western Union telegram. Each word was very expensive: something
on the order of $4 per word, which adds up pretty quickly for a poor,
starving graduate student. I remember spending long hours pondering
the three or four words I'd select for my telegram. Nouns and adjectives were discarded, too superfluous. I emphasized verbs and pro-
nouns. I'm sure that "LOVE YOU" and "MISS YOU" were among the
select phrases within the three or four telegrams I managed to send
during my four-month absence. The most important of these telegrams
consisted of four words: "LET'S MARRY THIS SUMMER." Absence,
combined with distance and isolation, forged me in great ways. And
so, following my first expedition to Antarctica, the course of my life
changed. Ferne and I were married a half year earlier than we originally planned.

In the mid-1980s, when I first began a series of nine consecutive research expeditions to McMurdo Station on the Ross Sea, we didn't have
anything as luxurious as the Internet. Twitter, Facebook, and YouTube
were figments of the imagination. Even telephones were only available
for life-and-death emergencies, and only the station leader could actually make the call. At the time, the U.S. Navy was providing logistical
support to the U.S. Antarctic Program, and this included air support
both on and off the sea ice runway at McMurdo Station, as well as "communications." The Navy's provision of this latter item included access
to a MARSGRAM. The "MARS" portion of this term stands for Military
Amateur Radio Service. Originally developed to raise the morale of
Navy personnel at sea, a MARSGRAM required the sender to write a

message on a form and drop it in the MARSGRAM box affixed to the wall of the station galley. Once every few days, Navy personnel collected the messages and sent them by radio teletype back to the United States, where they were printed and mailed via the U.S. Postal Service to their final destination.

Another communication option, equally unattractive, was a phone patch from the ham shack. When the atmospheric conditions were right, one could use a ham radio to make contact with a volunteer operator back in the United States, who would in turn call the phone number of a loved one and "patch" a call to them. Ferne and I did not like this arrangement. First of all, we had to say "over" when each of us had finished saying his or her piece, which allowed the fellow assisting with the patched call to hit a switch, permitting a response by the other party. This scenario also meant that someone was listening to our conversation. Rumors abound that some radio operators broadcasted phone patches over a PA system so others could listen in, unbeknownst to the speaking parties. A friend who had worked on a ship in Antarctica told me his ship's radio operator routinely broadcasted phone patches for entertainment to break up the monotony of life aboard ship. One has to pity the poor soul whose passionate conversation with his wife or girlfriend was piped over the loudspeakers.

My solution to the communication conundrum at McMurdo Station in the late 1980s and through the early 1990s was to make a reservation each week to use the radio phone housed in a little booth in the front office at Scott Base, our neighboring New Zealand Antarctic station. On the appointed day, I would dress up in all my cold-weather clothing and walk the two miles to Scott. The road wound east around the edge of Observation Hill, a key landmark topped with a large wooden cross honoring Robert Scott and his men—Edward Wilson, Henry Bowers, Lawrence Oates, and Edgar Evans—all lost in 1912 in unseasonably late summer storms on their return from the South Pole. Emerging from Ob Hill you can see Scott Base nestled on the edge of McMurdo Sound at the bottom of a long, gently descending snow-covered hill. With clear weather, one can catch a stunning view of Mount Terror to the northeast and Mount Erebus to the north, the latter being the only active volcano in Antarctica and towering 12,280 feet above our station. Across the frozen McMurdo Sound to the south lies Black Island, a landmark pointing in the direction of the Beardmore Glacier that Scott's party climbed, hand-hauling their heavy sleds on their fated journey to the South Pole.

I walked to keep warm, inhaling air so cold it hurt my chest, yet as clear and fresh as any on the planet, devoid of the diesel fumes that typified the air in bustling McMurdo Station. The depth of the silence and solitude was uncanny. As I dropped toward Scott Base, the road passed close to the edge of the frozen sea, and I often stopped to ponder the evenly spaced frozen waves comprised of long pressure ridges in the sea ice that stretched to the horizon. Having spent much of my youth surfing in Southern California near Santa Barbara, I couldn't help but think that this was a frozen version of the surfer's iconic vision of "endless waves."

If the weather was inclement, I could catch the shuttle bus to my phone reservation. It was normal for science support jobs at Antarctic research stations to be filled by people with surprising backgrounds, and as such there was always someone with an interesting story driving the shuttle. I have met PhDs in nuclear physics wiping down the tables in the galley, premedical students shoveling the ice and snow from under the buildings, and retired English teachers driving shuttle buses. Rumor has it that thirty Americans apply for each job available at McMurdo, Palmer, or the Pole. Many come to satisfy their curiosity about Antarctica, to pursue their side interests in natural history or the

Antarctic heroic era, or just to say that they've been there. Most fall in love with Antarctica and return time and again. Getting back to "the ice" in and of itself becomes a vocation.

During the austral springs of 1985 and 1986, I would sit in the overheated front office of Scott Base, growing impatient and waiting to hear my name called. Finally, my turn would come. Handing my home phone number over to an office assistant at the front desk, I would make my way to a cramped phone booth and pull the door shut behind me. When the phone rang, I would lift the receiver. "Hi Ferne!" After what seemed like an eternity, I would hear her faint response. She might as well have been speaking to me down a long hollow tube with a modicum of static thrown in for good measure. The long delay between our familiar exchanges was frustrating, and on occasion, humorous. One minute we would remain silent in anticipation of the other speaking, the next minute we would trip over one another midsentence. Our words were delayed by 2,400 miles of radio distance to New Zealand, and then 6,500 miles of telephone cable spanning the Pacific seafloor. But the privacy of the call sure beat the public Ham Radio phone patches. One year, magnetic radiation generated by sunspots interfered with our weekly phone calls. I walked over to Scott Base repeatedly, only to discover the phone down each time. Out of touch for several weeks, Ferne worried for my safety, and soon others, including the members of her choir, joined her growing concern. Finally, the sunspots dissipated, and I was able to call and explain my absence. When Ferne told her choir about the sunspots, they all burst into laughter. Sunspot excuse jokes have followed her to this day.

By the mid-1990s, McMurdo Station was poised to join the world of modern communication. With a communications satellite in polar orbit, it became possible to connect to the Internet and to use Internet Protocol (IP) phones. I remember that those administrators in charge of overseeing station operations were reluctant to open the station up to modern communication, especially email, fearing that computer games and chatting with friends online would interfere with daily tasks. The opposite turned out to be true. Email and IP phones instantly brought the adventure and excitement of living and working in Antarctica to family, friends, and especially the classroom. The ability to send reams of data to science laboratories around the globe was awesome. By 1997, an IP phone was available in every hall of every floor of every dormitory at McMurdo Station. Calls were easy and affordable. We were no longer isolated.

When I arrived at Palmer Station in 2000, the station still had limited access to the Internet. IP phones were not yet available. Instead, we could call home periodically via satellite phone. Cumbersome and large by today's standards, the satellite phone had a ten-inch collapsible antenna that clicked into an upright, outstretched position. Because the phone relied on an uninterrupted line of sight to secure a connection with a satellite, the user had to stand outdoors to make a call. Using these devices is inherently awkward because the phone could only lock on to a satellite for a few minutes at a time before it moved out of range. It was then a matter of redialing and waiting. One night, in the middle of a blizzard, I didn't want to miss a promised call with Ferne, so I braved the elements and stood outside the BioLab at Palmer Station. The wind howled and drove stinging snow crystals against my face. I hunkered down, back to the wind, and pointed the satellite phone toward the stormy night sky. By the time we finished our call it felt like mild hypothermia had set in.

Phone communications at Palmer have since taken a huge turn for the better. An IP phone sits on my office desk in the laboratory and another rests on the table next to my bed. I can also check out a portable phone for the entire field season. This portable phone operates as an

IP phone, but allows me the luxury and the security of having a phone in my pocket, much like a cell phone. Recently, I called home from the summit of the Marr Glacier behind our station, surrounded by jaw-dropping vistas and the glittering icy sea framed by the snow-covered peaks that are the southern extension of the mountain chain that forms the Andes. Times have changed indeed.

# Chapter 2

## It Is All about the Ice

*I* *have grown accustomed to the quirks of sailing south across the* Drake Passage, and I've come to expect the anomalous temporal and spatial qualities of air and sea temperatures, winds, ocean currents, icebergs, sea ice, ice sheets, and glaciers whose synergy defines Antarctica's physical essence. The interplay of these physical attributes becomes an ever-changing stage upon which life performs. The ecological impacts of rapid climate change on the marine life of the Antarctic Peninsula are inseparably linked with the environment and geography. Neither physical nor biological realms exist in a vacuum; rather, the two collectively form a complex web of interactions, some relatively benign and others triggering cascades of ecological consequences. In 2009, I led an Antarctic Climate Change Challenge Mission Cruise for Abercrombie and Kent Travel Company to the western Antarctic Peninsula. Among the participants from my hometown in Alabama were four undergraduate honors students from the University of Alabama at Birmingham. Each had enrolled in my Antarctic Marine Ecology course, and between scholarships, student loans, and artfully approached parents, each student had managed to scrape together the necessary funds for the voyage. Of the remaining two hundred passengers aboard our ship, the *Minerva,* most were well-heeled, retired professionals who had traveled the world over, and they shared my sense that these four students were incredibly fortunate to visit Antarctica at such a tender age.

The first physical attribute of the Antarctic environment that I share with those new to the continent is the abrupt drop in seawater temperature that welcomes one to the Antarctic Circumpolar Current

(ACC). Circling the entire continent of Antarctica in a clockwise direction, the ACC is truly one of the world's greatest geographic wonders. Unparalleled in its volumetric dimensions (the current extends from the sea surface to depths ranging from six thousand to twelve thousand feet), its very presence ensures the refrigeration of Antarctica. This refrigeration is caused by the ACC trapping very cold, less saline surface waters that circulate around the continent and maintain stable cold air temperatures. When the ACC was formed many millions of years ago, the entire continent rapidly cooled. Even more remarkable is its immense breadth, extending at some points as wide as twelve hundred miles. Because the ACC circles the globe (the Southern Ocean is the only ocean to do so), it provides a unifying link to every major ocean basin. Some scientists refer to it as the "mixmaster" because exchanges of water masses across these linkages play a critical role in regulating the climate of our planet.

Strong westerly winds drive the current's eastward motion and have given the ACC its alias: the West Wind Drift. These westerlies are world renowned for their speeds. As such, some are named by combining catchy adjectives with latitudes; examples include the Roaring Forties, the Furious Fifties, and the Screaming Sixties. Such wind-intensified latitudinal gradations are the result of air being displaced from the equator toward Antarctica. As air travels south of thirty degrees latitude, it gradually loses heat garnered over the equator, and as it cools, it sinks closer to the earth's surface while the earth's rotation drives the air from west to east. The growing intensity of winds south of forty degrees latitude is the result of a vast expanse of uninterrupted ocean that permits the winds to build upon themselves relentlessly. I know these winds all too well. I was essentially baptized by wind on my first Antarctic expedition to Kerguelen. In the Northern Hemisphere, land masses interrupt the wind's flow, preventing it from building to great speeds.

One of the projects I assigned the students on the *Minerva* was to locate and identify the ACC by recording surface seawater temperatures and positional coordinates every three hours over the thirty-six-hour Drake Passage crossing. It didn't take the students long to discover a GPS on the ship's bridge that provided our latitude and longitude and also a digital gauge that displayed surface seawater temperature. Determining who would set his or her alarm and get up repeatedly throughout the course of the night to record GPS coordinates and sea temperatures was even more challenging. Halfway across the Passage, my students eagerly approached me at dinner and presented a graph that displayed seawater temperature and latitude. The students explained that as we had approached 57 degrees latitude, the seawater temperatures had transitioned in just six hours from a moderate 46 degrees to a more frigid 38 degrees Fahrenheit. Here, they exclaimed, was irrefutable evidence we had crossed the Polar Front and entered the ACC.

Scientists predict that as rapid climate change increasingly impacts the Southern Ocean, the ACC will act as a conduit to transmit climate changes around the globe. A scientist at the Scripps Institute of Oceanography reported in a groundbreaking paper in the journal *Science* that since the 1950s the average mid-depth temperature of the Southern Ocean (especially within the ACC) has significantly warmed—and done so at a faster rate than temperatures measured in the Pacific, Atlantic, or Indian Oceans.[1] The implications of this Southern Ocean warming trend are profound as the waters surrounding Antarctica spread their influence gradually over the entire globe. Importantly, observations indicate that the ACC has migrated about thirty miles toward the South Pole since the 1950s. In another paper, scientists from the Canadian Center for Climate Modeling and Analysis[2] put this southern migration of the ACC into perspective by using oceanographic models. They concluded that by the end of the twenty-first century the migration (actually, shrinkage) of the ACC could displace a body of water equal to that of the entire Arctic Ocean. As such, the climate influence of a diminished ACC would be felt globally. Marine scientists at the Rosenstiel School of Marine and Atmospheric Science at the University of Miami point out that as the westerlies and ACC continue to move south, the oceanic "gateway" at the southern tip of Africa will expand as warm salty waters from the Indian Ocean leak into the Atlantic Ocean.[3] This enhanced leakage influences patterns of oceanic circulation. Specifically, warm saltwater may disrupt currents climate scientists had predicted would reduce warming in the North Atlantic Ocean.

If my students were to travel south across the Drake Passage in a future world—a world where the ACC and its nutrient-rich water and abundant seabird population had shrunk and was located closer to Antarctica as a result—their second assignment would have been abbreviated. I assigned my students to spend as many daylight hours as possible out on the rear deck recording the numbers and types of seabirds following our ship. When the students reviewed their avian field notes, they discovered that as the *Minerva* had transitioned into the colder, nutrient-rich waters of the ACC, both the abundance and diversity of seabirds had dramatically increased. Black-browed alba-tross had been joined by their cousins, great wandering albatross and royal albatross; cape petrels now enjoyed the company of giant petrels, sooty shearwaters, and Wilson's storm petrels. The students correctly surmised that increased quantities of avian prey—zooplankton and fish—explained this transition. Off the rear deck, seabirds soar on the westerly winds against a backdrop of sea and sky. Most circle in pat-terns that take them gradually up the sides of the ship, then out in front of the bow, and with sudden angling of wings, back at breakneck speed to once again soar behind the ship's stern. This pattern of flight pro-vided a perfect opportunity for my students to observe the phenomena of *dynamic soaring*—a flight technique that exploits the ability of birds to gain speed by crossing back and forth between calm and windy air masses. Seabirds dive into the calm air behind a ship or wave and then wheel back up and into a strong headwind. When they emerge like this, the speed of the air flowing across the top of their wings increases—then by shifting their flight to a downwind glide, the birds are propelled, like a high-speed roller coaster over a drop, into yet another calm region. This dynamic process provides an almost effortless means of travel and makes possible such feats as the transoceanic foraging flights of the giant wandering albatross.

No voyage from South America across the Drake Passage to Antarctica is complete without celebrating the first sighting of an iceberg. Usually one can expect to see one about two-thirds of the way across the Passage. On Antarctic cruise ships, a bottle of fine champagne is awarded to the first guest to inform the officer on the bridge of the sighting. Aboard research vessels, scientists are outwardly subdued, as if sighting the first iceberg is routine and benign. But scientists hide their excitement behind their professional demeanor. Icebergs are irrefutably stunning—transcending both science and art. When sunlit on a clear day, they are so brilliantly white they are impossible to look at without squinting. Where snow has melted or blown free, a translucent-light to deep-azure blue emerges from the ice. At the water line and up to twenty or so feet below the sea's surface, the shades of turquoise and lime green can take an onlooker's breath away. As icebergs melt and shrink, they periodically tip over, exposing their underbellies of glassy-smooth or pockmarked surfaces. As a precaution against sudden rollovers, our vessels give icebergs a wide berth, and the National Science Foundation (NSF) Antarctic dive program prohibits scuba diving in their proximity. But one need not enter the water to enjoy such surreal beauty. One of my favorite places in Antarctica to take nature photographs is from a small boat slowly winding its way through gardens of icebergs. As if sculpted by an artist, their myriad shapes—some reminiscent of animals, castles, or treasures from the world's finest modern art collections—provide a virtual smorgasbord of photographic potential.

Icebergs are large pieces of floating freshwater ice, generally projecting from just a few to around two hundred fifty feet above sea level, and weighing hundreds to millions and occasionally even billions of tons. They can originate from a variety of sources, such as being shed from a snow-formed glacier (known as a *calving event*), or from cracking off an ice sheet. The word *iceberg* itself is derived from the Dutch *ijsberg*, meaning "ice mountain." This term is somewhat ironic considering that 80 to 90 percent of an iceberg is hidden below the surface of the water. If upside down, however, an iceberg would truly be an ice mountain.

Antarctica boasts some of the largest icebergs ever recorded. In 2000, Iceberg B-15 broke free from the Ross Ice Shelf—a floating platform of ice that can be hundreds or even several thousands of feet thick. (Large icebergs are given designations so they can be tracked.) The Ross Ice Shelf is the largest ice shelf in Antarctica and is about the size of France. Remarkably, a vertical ice wall towering 50 to 100 feet fronts 370 miles of the Ross Sea. The ice shelf was named for Captain James Ross Clark who first sighted the shelf on January 28, 1841. Iceberg B-15 measured twenty-two miles wide and 183 miles long, and was estimated to weigh three billion tons. At 4,200 square miles, the iceberg was 8 times the size of the city of Los Angeles. I was fortunate to witness one of these behemoths on a cruise to Antarctica aboard the *Explorer II*—on which I was lecturing with a friend, geologist Henry Pollack—in 2007. We sailed for hours, seemingly within reach of a thirty-one-mile-long iceberg that had grounded itself near Clarence Island off the northern tip of the Antarctic Peninsula. Henry, who had visited Antarctica frequently over a span of eighteen years, was, like me, nonetheless awestruck by this iceberg's grandeur. Emerging one hundred vertical feet from the sea, the immense iceberg towered above us, dwarfing our ship as we passed.

As global temperatures rise, icebergs will more often break off, or calve from, the mainland. Throughout the decade I have worked at Palmer Station, I have witnessed many bergs or smaller pieces of ice calve from their glaciers. About once a week, I would be startled by a loud, thundering crash. Leaping from my desk on the second floor of the Palmer BioLab, I would join others running down the hall to throw open the door and watch the waves rolling up neighboring Arthur Harbor— waves brought on by a house-sized chunk of the Marr Glacier breaking free and plummeting into the sea. Now when I visit Palmer Station, the calving events have become so routine that my colleagues and I in the BioLab don't even bother to move from our desks when we hear the glacier roar. Sometimes, three or four calvings happen in a single day. Indeed, those who have worked at Palmer Station over the past decade don't need to consult journals, television programs, or the Internet to understand how the climate is changing. Furthermore, as the geography changes, so do the names of actual locations. When the receding ice tongue of the Marr Glacier recently revealed an island rather than a seemingly long-established point of land, Amsler Island was officially born.

The frequency of icebergs calving off glaciers and ice sheets breaking up will inevitably increase as waters along the Antarctic Peninsula and other regions of western Antarctica continue to warm. Larger icebergs will also become more common as they are shed from ice sheet break-ups, and their increased mass will permit them to drift farther north before finally melting. They will also begin to show up in odd places. On November 16, 2006, while I was on sabbatical at the University of Otago in the city of Dunedin on the South Island of New Zealand, government officials spotted a large iceberg off the coast—the first iceberg sighted from a New Zealand shoreline in seventy-five years. It turned out to be one of a flotilla of over a hundred icebergs. Immediately, the media was abuzz with news about the impacts of climate change, and the New Zealand icebergs soon became a major tourist attraction. Tourists paid $300 apiece for a helicopter ride over the fields of ice, and for a bit more cash, they could land on the large berg within sight of Dunedin. A tour company quickly organized an iceberg wedding, but had to scuttle it when the helicopter pilot decided it was unsafe to land.

A warming Antarctic Peninsula riddled with icebergs has consequences that are hidden from the casual observer. The increased amount and sizes of icebergs scouring the coastal seafloor disrupt the marine communities there. Marine biologists have long known that near-grounded icebergs behave much like earth movers at construction sites, displacing tens of thousands of square yards of seafloor sediment. The exposed portion of an iceberg acts as a sail, transferring the energy of wind and current to motion, causing the berg's base to plow through soft sediments and scrape over hard bottoms. Massive iceberg scars extend for miles along coastal Antarctic seafloors, and these are devoid of seaweed, sponges, sea anemones, soft corals, sea spiders, starfish, brittle stars, and even fish. Over a period of a few years, the process ecologists refer to as *community succession* will kick into gear along these iceberg scars. Temporary communities of rapidly settling and fast-growing, but short-lived, seaweeds, sponges and sea squirts will give way to more stable "climax" communities comprised of more competitive seaweeds and marine invertebrates that grow slower and have longer life spans. Climax communities are ecological communities in which populations of bacteria, plants, and animals remain stable and exist in balance with each other and their environment. In the big picture, Antarctic seafloors that are subject to intermediate levels of periodic iceberg scour are checkered with short-term opportunistic and long-term stable communities and, as such, sustain higher overall diversities of species. But just as intermediate levels of iceberg disturbance may increase species diversity, too much iceberg disturbance may actually compromise this diversity. Heavily ice-scoured seafloors, like a graded construction site, can be biological deserts. Climate warming could result in an overabundance of coastal icebergs that regionally decimate near-shore seafloor communities.

Icebergs can also have unforeseen impacts on Antarctic marine birds. Following the 2005 break-up of B-15, a massive offspring (renamed B-15A) grounded at the mouth of McMurdo Sound. Its position effectively blocked the outflow of pack ice from the Sound while simultaneously cutting off the Adélie and emperor penguins from their food resources. This blockage diverted the penguins to a route that effectively doubled the distance they would normally travel, from about 60 to 120 miles. Until the massive B-15A iceberg floated free several years later, biologists routinely found emaciated penguin carcasses en route between rookery and sea.

*Having completed our crossing* of the Drake Passage, my students and I rendezvoused on the ship's bridge and took in the jaw-dropping vistas of icebergs that ranged in size from a Volkswagen Beetle to a large hotel. "Are all icebergs so easy to see?" one of my students asked the ship's captain. The captain explained that the crew manning the bridge can sight small icebergs emerging above sea level (aided by a powerful spotlight at night), and that the ship's radar readily detects the larger icebergs. He further explained that they had to give the large bergs a wide berth because their ice can extend horizontally just below the sea's surface—a significant hazard to ships.

Unfortunately, technology has yet to provide a means of detecting smaller icebergs whose upper portions lurk just barely above the sea surface. Known as *growlers* because of the growling sound they emit as air escapes the ice, these are about the size of a concert grand piano (fifty square feet). Because they have lost most of the air trapped in their ice, they sink low into the water, often protruding only two to three feet above the sea surface. Growlers are too small to be detected by radar and too low on the horizon to be seen from a ship's bridge, yet they're large enough to breach a ship's hull. Ships that routinely ply Antarctic waters are generally designed with ice-hardened (thicker-than-normal) hulls, or even double hulls to prevent a breach should a collision occur. The NSF research ships I sail upon to Antarctica and the small Antarctic tour ships leased by Abercrombie and Kent have ice-hardened hulls. But not all ships that visit Antarctica have reinforced hulls, including the larger cruise ships that carry several thousand tourists at a time. This is a cause for concern, especially as a rapid warming along the Antarctic Peninsula increases the frequency of encounters between ships and icebergs. Over the past twenty years several of the smaller Antarctic cruise ships such as the *M/S Explorer* have had to evacuate their passengers at sea. The *Explorer* collided with an iceberg on November 23, 2007, and sank to the seafloor within fourteen hours. Fortunately, the weather was calm and the passengers and crew were plucked from their lifeboats by a cruise ship that happened to be nearby. But the prospect of evacuating several thousand passengers in heavy seas and foul weather, with assistance perhaps many hours away, portends a disaster of *Titanic* proportions.

Scientists attempting to measure how rapidly the climate along the Antarctic Peninsula is changing are fortunate because they can access Faraday Station (now Vernadsky Station), a small coastal research facility, some twenty-five miles south of Palmer Station on the central western Antarctic Peninsula. The British government established Faraday Station

in 1947, where scientists initiated and sustained air-temperature measurements that span a period of six decades. When a scientist plots the
mid-winter air temperatures over this period on a line graph, he or she
sees an average increase of almost 2 degrees Fahrenheit per decade.[4]

Sea temperatures along the western Antarctic Peninsula are also
warming. Scientists have determined from seawater temperature measurements taken from 1955 to 1998 that a 1 to 2 degree Fahrenheit
increase has already occurred in surface seawater temperatures.[5] And
increased sea and air temperatures are contributing to the rapid retreat
and decreased duration of the annual sea ice (ice that accumulates and
actually doubles the size of Antarctica every winter), which, over the
past thirty years, has seen a reduction of 40 percent along the Peninsula.
Some scientists predict that by the middle to end of this century, annual
sea ice along the western Antarctic Peninsula will become scarce.[6]

All these factors make the central western Antarctic Peninsula the
poster child for climate warming on the planet. While climate scientists originally thought that regions of the Antarctic continent other than
the Antarctic Peninsula were less vulnerable to warming, this view has
changed dramatically over recent years. Air temperatures along the
west coast of the Antarctic continent have been warming according to
a recent study published in *Nature*.[7] Using a combination of satellite
thermal infrared observations and ground weather stations, the authors
of the study have determined that west Antarctica has warmed at a rate
exceeding 0.18 degrees Fahrenheit per decade over the past fifty years.
These findings were a game changer, as one of the outstanding questions in Antarctic climatology has been whether the dramatic warming
on the Antarctic Peninsula had also been going on in West Antarctica.
The study answered this question with a resounding "yes." Polar climate
scientists attribute the rapid warming of the Antarctic Peninsula to a
combination of the buildup of greenhouse gases and, especially during

summer months, changes in wind patterns (the westerly jet) due to the hole in the ozone. The authors of the *Nature* paper surmise that West Antarctica seems less influenced by ozone-related changes such as wind patterns and more tied to broad patterns of atmospheric circulation associated with regional changes in sea ice.[8] One intriguing hypothesis to explain why some areas of eastern Antarctica have not warmed is that altered wind patterns generated by the hole in the ozone have postponed warming.[9] Should the ozone hole eventually close, as now predicted by atmospheric scientists, warming in the eastern regions of the continent may make up for lost time. Regardless, Antarctica is warming, and the Antarctic Peninsula is at the forefront.[10] From a scientific standpoint, the dramatic warming of the Peninsula provides an opportunity to study first impacts of unprecedented warming on sea ice, icebergs, ice sheets, and marine life.

Iceberg-studded seas give way to first views of the Antarctic Peninsula and its neighboring islands. Glaciers that extend to the sea from thick land-based Antarctic ice sheets adjoin semi-permanent ice shelves that float on the surface of the sea. Some of the ice shelves along the east and west coasts of the Antarctic Peninsula are mind-boggling in their dimensions. Given the rapid warming of air and sea temperatures over the last thirty years, large sections of at least nine ice shelves have separated from the Peninsula, most of these since 1995. Perhaps the most famous break-up to date is the Larsen Ice Shelf located on the eastern side of the central Antarctic Peninsula. On January 31, 2002, satellite images revealed large numbers of striations running shore to sea in a region known as Larsen Ice Shelf-B. A little over three weeks later, a satellite image on February 23 showed that the striations had led to disintegration, as tiny to massive iceberg-sized chunks now littered the region. Two weeks later, on March 5, a satellite image indicated that Larsen Ice Shelf-B was essentially gone; in its place floated hundreds, even thousands, of icebergs in the Southern Ocean. The magnitude of this event is impossible to put to scale from satellite images alone— without a scale, it may seem like mere tens of miles of ice were affected. But scientists have estimated that the vast region of Larsen Ice Shelf-B that broke out measured a staggering 3,540 square miles, a piece of real estate about the size of Rhode Island.

The breakout of this major section of the Larsen Ice Shelf provided the stimulus for the LARISSA Project. This multinational research program, funded by the U.S. National Science Foundation, involves twenty-three principal collaborators and investigators teamed together to address the dramatic environmental shifts in Antarctica's Larsen Ice Shelf system. In the big picture, the disintegration of this ice sheet is a significant regional problem with global implications. Amy Leventer, a geologist, friend, and colleague from my early days as a polar marine biologist at McMurdo Station in the mid-1980s, was among a group of researchers joining the LARISSA Project from widely varying scientific backgrounds. In addition to Amy, glaciologists, physical and biological oceanographers, marine ecologists, and climate and atmospheric scientists were present. Each individual brought to the table a different set of attributes that, when combined, generated a compelling synergy. As complex problems such as global climate change increasingly challenge humankind, broad interdisciplinary science teams such as those exemplified by the LARISSA Project provide an optimal approach to answering complex questions.

Amy explained to me that the LARISSA Project undertook yearly research cruises to the Larsen Ice Shelf aboard the *Nathaniel Palmer*. The National Science Foundation provided additional logistical support, including the use of two helicopters and a Twin Otter airplane. Each year, once their ship had arrived in the vicinity of the Larsen Ice Shelf, the LARISSA scientists were deployed by small boats, helicopters, or the

Twin Otter to sample seawater and ice and to dredge, core, and record video of the seafloor communities under the former and current ice shelf, and to establish a suite of automated battery-powered monitoring stations both on the surface of the ice shelf and adjoining glaciers as well as on bare rock near glaciers feeding into the ice shelf. The ice monitoring stations, known as *AMIGOS Stations*, each consisted of a GPS unit that provides precise information on the position of the ice shelf or glacier, digital instruments to collect weather information (temperature, humidity, and wind speed), and a computer modem that can store and download data directly in real time to an orbiting satellite. With sufficient ice monitoring stations positioned in key locations, the Larsen Ice Shelf and its adjoining glaciers are like a living organism equipped with sensors to monitor its respiration and heartbeat. Capturing a shelf breakout in real time through these devices would be akin to monitoring a heart attack while it was in progress; scientists could analyze the intricate dynamics of the ice shelf as it broke out, much like doctors in the emergency room use information on the condition of a damaged heart to predict future attacks. As a geologist, Amy was personally involved in establishing a number of the bedrock monitoring stations that scientists deployed on ice-free rock surfaces near the glaciers.

LARISSA scientists also provisioned the monitoring stations with sensitive instruments that record the earth's surface actually rising, or rebounding, in response to the reduction of the weight of the glacial ice. As such, scientists can precisely measure what's known as the earth's *glacial isostatic adjustment.* To envision the earth's rebound, think of the surface of hardened gelatin. When you push on the surface with your finger, it depresses. When you remove your finger, the gelatin readjusts to its original position. Scientists measuring the earth's isostatic adjustment can use this information to determine how Antarctic glaciers will ultimately contribute to global sea level rise.

In January 2010, with Amy and her fellow LARISSA scientists aboard, the *Nathaniel Palmer* departed Punta Arenas, Chile, for the Larsen Ice Shelf. Unfortunately, as the *Nathaniel Palmer* rounded the northern tip of the Peninsula, the captain observed that the annual sea ice had not broken out and the ship would be unable to reach the shelf. The helicopters were useless over the distance from the ship to the Larsen Ice Shelf. Amy summarized the collective feeling of those on board: "The ship, the helicopters, the plane, and the scientists were all dressed up with nowhere to go." They could either return to Chile empty-handed or find another way to reach the ice shelf. Using his ingenuity, the captain repositioned the ship so that the Larsen Ice Shelf was a direct shot to the east across the breadth of the center of the peninsula. The helicopters were now close enough to the ice that they wouldn't run out of fuel. However, they still had to deal with the impressive extension of the Andes that jutted in the middle of the flight path. Most of the peaks are steep, windswept, and extremely rugged, with elevations averaging above 6,000 feet—the highest peak is an impressive 9,186 feet. Amy and her research colleague Gene Domack, a fellow geologist from Hamilton College in New York, loaded their survival gear, food and water, and the modular components of an automated bedrock monitoring station onto one of the helicopters. Amy was comfortable around helicopters—she had flown hundreds of helicopter missions during her years of geological research in the vicinity of McMurdo Station. She even married one of the helo pilots stationed there. Despite all that experience, as Amy and Gene's helicopter lifted off the flight deck of the *Palmer*, she couldn't help but feel a bit nervous about her first flight over the mountainous spine of the central Antarctic Peninsula.

Like Amy, I have flown in helicopters among the stupendous Antarctic Dry Valleys—remarkable landscapes free of snow and ice due to lack of precipitation and high wind scour—that stretch for miles from the Transantarctic Mountains of Victoria Land to the coast of McMurdo Sound. On a crystal-clear austral summer day, flying by helicopter down the gut-dropping length of the Taylor Dry Valley and erupting out of its mouth over the deep-blue waters set against the sparkling white expanse of McMurdo Sound's ice edge is breathtaking. Add in the view of 12,280-foot Mount Erebus with its perennial smoky summit and you have a view Ansel Adams would have made famous in a photograph. How could any panorama on earth be as compelling? Yet Amy described her cross-peninsular flight that day as even greater in grandeur: "Dwarfed by the sheer magnitude of the landscape, our helicopter would climb and climb, only to emerge over seeming summits to reveal yet another even higher ridge to ascend." Amy had never felt as far removed from humanity nor as geographically isolated as the moment she and Gene watched their helicopter depart. As the thumping of the rotor blades faded into the distance, they stood alone in the deafening silence of Antarctica on a patch of bare rock near a glacier adjoining the Larsen Ice Shelf. With nary a research vessel or cruise ship along the entire eastern Peninsula, they were as alone as a human can be barring space travel. With little time to waste, Amy and Gene set about their day-long task of erecting the modular bedrock monitoring station.

The most recent breakup of a major ice shelf along the Antarctic Peninsula was observed March 25, 2008, when a 156-square-mile chunk of the Wilkins Ice Shelf broke apart. In what amounted to the first winter breakout in May 2008, another section disintegrated that measured 62 square miles, leaving the entire Wilkins Ice Shelf dangling from a cluster of islands by a thread of ice (only 1,640 feet at its narrowest point). When the strip eventually breaks it will release the entire ice shelf—a whopping 8,700-square-mile chunk of real estate the size of Connecticut.[11] Adding to the problem is the recent discovery by scientists from the Lamont-Doherty Earth Observatory and the British Antarctic Survey that strengthening currents are bringing more deep, warm water into contact with the underside of the nearby half-mile-thick Pine Island Ice Shelf, causing it to melt.[12] Nineteen cubic miles of the ice shelf's underside melted in 2009 alone, accelerating the flow of ice into the sea from the Pine Island Glacier that feeds into the ice shelf. Since 1974, the Pine Island Glacier has accelerated its flow into the sea by more than 70 percent, and thinned five feet per year between 1992 and 1999.[13] Given the rapid warming that has occurred along the length of the Antarctic Peninsula,[14] one could reasonably conclude that the ongoing disintegration of the Wilkins Ice Shelf is tied to climate warming, an observation that should serve as a warning that the coastal ice sheets lining the western coastal regions of the Antarctic continent may follow suit. Recent satellite-based measurements showing thinning of land-based ice in the western coastal region of the continent suggest that the increased disintegration of sea-born ice shelves along the Antarctic Peninsula may transition farther south. Furthermore, glaciologists estimate that if the land-based ice sheet that covers western Antarctica were to melt, it would raise global sea levels by about ten feet.[15] In such a scenario, Manhattan would be underwater and coastal Florida would be history.

If a silver lining exists on this rash of ice shelf breakouts occurring up and down the length of the Antarctic Peninsula, it is that these ice shelves rest on the surface of the sea. As such, when icebergs break off and drift north, melting along the way, they have no impact on global sea level. The reason for the lack of sea level rise is simple and can be easily illustrated with a glass of ice water. Ice that melts in the glass does so without changing the water level—the volume is consistent. The same principle applies to Antarctica's melting ice shelves: they will not affect the global sea level.

The loss of ice shelves, however, does produce a significant side effect. Ice shelves act as barriers that block land-based glaciers from flowing into the sea. Remove the barrier, and the glaciers flow more rapidly.[16] This increase of glacial ice introduced to the sea contributes directly to increasing global sea levels. Ted Scambos, a LARISSA project lead scientist and glaciologist from the National Snow and Ice Data Center at the University of Colorado, explained to me that when ice shelves are removed it is similar to removing a dam that blocks the flow of water down a stream. Rates of unimpeded glacial ice flow may increase by two to four times the rate of glaciers that interface with intact ice shelves. In the 2007 Intergovernmental Panel on Climate Change report, the scientific committee responsible for writing up the document chose to leave out any estimates of land-based glacial and ice sheet melt for Antarctica or Greenland in their predictions of global sea level rise.[17] At the time, they felt that the information would be premature. However, with new knowledge about the alarming decay of ice in both Antarctica and Greenland,[18] little doubt remains that the IPCC's next report (due out in 2014) will increase earlier predictions for global sea level rise by mid-century to its end.[19]

Scientists and support staff living at Palmer Station know their glaciers well. The Marr Glacier, towering behind the station, is by far the most prominent geographical feature in the region. As the glacier melts, its sea-side flanks on either side of the rocky point of land housing our station shed iceberg-sized chunks of real estate, while behind the station the glacial tongue recedes across the bare ground like a garden snail pulling its foot and head back into its shell. The Marr Glacier provides us with a visual barometer of our earth's rapid warming.

Maggie Amsler, a research associate in my laboratory, and a pioneer among women scientists in Antarctica, has an even more profound relationship with the Marr Glacier. At the crack of dawn, while the rest of the station still sleeps, Maggie extricates herself from a sleeping bag within her bivy sack (short for "bivouac sack," a thin fabric shell that fits over a sleeping bag) in the boulder-strewn landscape behind Palmer Station where she spends most nights. Depending on the weather and snow conditions, she heads off to hike, run, or ski to the summit of the glacier and comes back before breakfast. Her summit forays are mythic, but verifiable as I routinely encounter Maggie in the BioLab dining hall fresh from an early morning glacial assault. In 1979, Maggie arrived to spend the first of many field seasons at Palmer Station. She tells me that she used to be able to open the back door of the station and almost step on to the Marr Glacier. Today, it takes a hike of a third of a mile to reach the glacier's base. With findings that sync with the recessive behavior of the Marr Glacier, glaciologists have now determined that 87 percent of the glaciers along the Antarctic Peninsula are under retreat. Scientists have evaluated its retreat with aerial photographs of the glacier beginning in 1963, and then sporadically over subsequent years. The photographs show a dramatic pattern of glacial recession. Over recent years, improvements in technology have facilitated more precise measurements of the glacier's receding edge.[20]

Brian Nelson, the science technician at Palmer Station in 2010, has a full plate of science projects that he maintains in the TeraLab, a small building perched on the hill behind the main station. These projects include monitoring studies of seismic activity, air quality, ultraviolet radiation, and global measurements of lightning strikes. And there is one more scientific project that requires an annual outdoor adventure. Each year, Brian hikes along the receding edge of the Marr Glacier to map its exact position. By doing so, the rate of the recession of the glacier can be mapped over years and decades. Taking advantage of orbiting satellites and a GPS base station in the TeraLab, Brian can map the receding edge of the glacier to an accuracy of one centimeter. Brian explained to me that the high-tech Trimble surveying GPS unit, GPS receiver, and battery that he carries in his bright orange backpack are incredibly heavy to lug around. Brian operates the receiver using a handheld control unit wired to the GPS, much like a computer game. Brian attaches a circular dish antenna, about the size of a teacup saucer, to the frame of the back-pack on a pole extending several feet high. Each year, as the science tech sets out from the station across the rocky backyard wearing his gadget-laden backpack, fellow staff members mock him good-naturedly with comments like, "Hey, Ghostbuster!" Brian explained that despite the unsteady footing at the far ends of the ice, walking the glacier's terminus is fairly easy in its central region. At the south end, near Hero Inlet, the winter's snowdrifts persist well into the summer, and the snow is undercut by the water in the inlet below. Brian is at constant risk of punching his foot right through the snow and falling directly into the freezing water. Like a tightrope walker, he carefully tests each footing before applying his body weight. On the north end, the walk is even more exhilarating as the retreating glacier becomes quite steep, eventually forming a near-vertical wall whose base ends in a slippery muddy slope. To properly survey the glacier's edge in this area, Brian has to sidle up as close to the ice wall as possible before sliding back down the muddy slope then scrambling back up to the base of the wall to repeat the process. In addition to the danger of large pieces of ice falling from the wall, Brian is periodically vulnerable to downbursts of small shards of ice. Once, he nearly lost a shoe where small streams of glacial melt-water had turned the fine-grained soil into a sloppy quicksand. After a couple of hours, Brian returns to the station and downloads the GPS data to a computer. With a few keystrokes, he adds another year's profile to a historical map of the Marr Glacier that tells a tale of glacial retreat as rapid as anywhere on the planet.

One of my first encounters with an Antarctic glacier was in the Taylor Valley, a desolate but beautiful dry basin in the Transantarctic Mountains of Victoria Land some fifty miles west of McMurdo Station.

Our helicopter pilot took us on a brief tour of the eighteen-mile valley to see the receding tongue of the glacier before he dropped us off at the New Harbor tent camp. During my stay at New Harbor, I wandered up the Taylor Valley, admiring the grandeur of the landscape and the glistening polished stones sculpted by the glacier and dust and sand blown by the notorious Antarctic katabatic winds. The term *katabatic* is derived from Greek and means "going downhill." Katabatic winds increase due to gravity and are comprised of dense air flowing from high to low elevation. These winds build to blistering speeds as they descend. In Antarctica, the katabatic winds plummet off the high ice sheets of the polar plateau down snow- and ice-covered terrain, glaciers, and dry valleys, whipping fine sand and dry snow into stinging blizzards and peeling the very surface of the sea skyward. British Antarctic heroic-era (the period from 1897 to 1922 marked by intense scientific and geographic exploration of Antarctica) scientists Edgeworth David and Raymond Priestley write that the steep grade in slope and decrease in temperature from the Polar Plateau to the sea and the vast surrounding ocean combine "to make the Antarctic the home of winds of a violence and persistence without precedent in any other part of the world."[21] Yet katabatic winds are not restricted to cold places. For example, growing up in Santa Barbara, California, I grew accustomed to the hot dry Santa Ana katabatic winds that howled relentlessly down the canyons toward the Pacific Ocean in the fall and early winter. Caught in their spell, the pine trees behind my home in the coastal foothills would moan and howl and bend, and the occasional Southern Californian forest fire would become an inferno. Yet Santa Ana katabatics pale in comparison to those in Antarctica.

My friend Sid Bosch, a marine biologist at the State University of New York at Geneseo, and his research technician Tom Gast experienced the wrath of katabatic winds on an austral spring October day near Palmer

Station. The two researchers were in a zodiac boat—an oblong, sixteen-foot rubber boat built of a series of rigidly inflated tube-shaped sections that are pressurized just like automobile tires—in Arthur Harbor, towing plankton nets for marine invertebrate larvae. Zodiacs are the work-horses for marine scientists at Palmer Station. Importantly, given the unpredictability of Antarctic weather, the small boats are seaworthy in sloppy swells and they maneuver effectively through the basketball- to table-sized chunks of sea and glacial ice known collectively as brash ice. Tom and Sid received a sudden radio call from the communications dispatcher at Palmer Station alerting them that the winds at the station had shot up from a normal range of about five to ten knots to twenty-five knots (about thirty miles per hour), the wind speed at which all zodiacs are recalled, and that they should return to station ASAP. Out of sight of the station and on the leeward side of nearby Janice Island, Sid and Tom were only experiencing mild wind. Yet as the two researchers cleared the leeward side of the small island, they were blasted by the offshore winds. Sid later estimated that it had to be blowing at least forty knots.

The wind lifted the bow of Sid and Tom's zodiac, threatening to flip it. With Tom operating the engine at the helm, Sid threw his body as far forward as possible to weigh down the bow. Had wind been the only issue, Sid and Tom would have probably made it back to station. However, the brute strength of the sudden burst had broken out the fast ice that had been attached to the shoreline, sending a flotilla of large chunks in their direction. Sitting high on the bow, Sid pointed his outstretched arm toward gaps in the encroaching army of car-sized ice chunks. Yet each time Tom gunned the zodiac's engine to propel the boat through, the gap in the ice closed, forcing a hasty retreat.

Helpless against the wind and ice, the researchers realized that making it back to the station was not an option. The situation was made all the more stressful because both researchers knew that if they were blown out to sea, rescue was unlikely. (The station manager admitted as much, as he would not have risked additional station personnel by ordering a rescue.) Sid and Tom turned their full attention to a last-ditch effort to secure a landing on one of the small islands adjacent to Palmer Station. At the last minute, Tom was able to gun the zodiac through a gap in the sea ice and land at Torgerson Island, the last island that remained in their path before the open ocean. After Tom had yanked the outboard engine up into its locked position, the two frantic researchers dragged the zodiac up and on to the rocky shore of Torgerson and secured the bow line to a boulder. With no chance of returning to station until the winds dropped, Sid and Tom located the island's chest-high emergency blue survival barrel. They removed from the barrel a small cache of emergency supplies including a mountain tent, two sleeping bags, fuel, and a backpacking stove, a medical kit, and dried food and bottles of water. They secured a site that was somewhat protected and set up camp for the night. Ever since Palmer Station opened in 1968, survival caches have been maintained on the small neighboring islands. Presently, nine of the islands have blue survival barrels.

By first light, the wind had died down, and the sea ice had dispersed enough that Sid and Tom could attempt passage back to Palmer. The two sleep-deprived researchers packed up their camp, launched their zodiac, and returned the short distance to station. Contributing to the long history of human survival, Sid and Tom's adventure had allowed Antarctica to serve up yet another poignant reminder of its fickle nature. The explorer and geographer Paul Siple, who represented the Boy Scouts of America during the famous Antarctic expeditions led by Adm. Richard E. Byrd in 1928–30 and 1933–35, is likely to have been the first person to utter the scout motto, "Be prepared," on the Antarctic continent. The intent of this simple but poignant statement remains an essential ingredient of Antarctic survival.

I had a memorable encounter with katabatic winds while aboard a cruise ship that had planned to travel through the Lemaire Channel (also known as Kodak Alley), just southeast of Palmer Station. About halfway down the channel we suddenly encountered seventy-mile-per-hour hurricane-force gusts. Those of us on the ship's bridge watched a stunning display of nature's energy as the blistering winds lifted the upper inch of the sea's surface skyward to dissolve in a rainbow of spray and mist. Fortunately, our ship was headed directly into the prevailing wind and we were in little danger of being driven into the rocks on either side of the narrow channel. By the time we exited the Lemaire Channel an hour later, the winds had abated, but it wasn't the end of them.

Our cruise ship was equipped with twelve rubber zodiac boats that would be used to take guests to shore and back and closer to the dramatic scenery. I was in charge of operating one of the boats myself. The crane operator lifted the first zodiac off the top deck of the ship, hoisted it over the side, and brought it to an abrupt stop suspended about twenty-five feet above the water. From this position, a boat operator could hop into the suspended boat and hang on to the boat's ropes as it was then lowered by cable to the sea. As we watched, the first operator hopped aboard the dangling rubber boat and grasped the ropes attached to the deployment cable. At that very instant, a sudden burst of katabatic wind caught the zodiac in a ferocious gust, flipped it sideways like a child's toy, and effectively turned it into a kite. The woman who was in the boat now found herself dangling above the icy sea, clinging to the ropes for dear life. Fortunately, an attentive ship's crew jumped to the boat operator's assistance, and in the blink of an eye she was standing back among us, dazed but unharmed. As the katabatic winds continued to blast down the adjacent valley, the small boat was hoisted back to its stowed position on the ship's upper deck, and tour operations wisely aborted for the day.

❦

*The physical attributes of the Antarctic Peninsula*—its icebergs, annual sea ice, ice shelves, glaciers, winds, and currents—are important players in a rapidly warming environment. Some are increasing in size and abundance (icebergs), some are diminishing in duration, size, or extent (annual sea ice, ice shelves, and glaciers), and others (winds and currents) are subject to regional variation. All are subject to change. When considered collectively, they portray a dynamic ecosystem undergoing remarkable transition in a relatively short period of time. These incredible changes affect the myriad of Antarctic marine organisms that over the millennia have adapted to survive in one of the world's most stable locations. Some of these organisms may adapt, but the vast majority of species here have become so finely tuned to their surroundings that they don't have much wiggle room.

# Chapter 3

# *Life Adrift*

## THE SMALL ORGANISMS MATTER

*E*arly naturalists surmised that the icy seas of Antarctica would be inhospitable to life. So when scientific exploration began in earnest at the dawn of the twentieth century, scientists delighted in their discovery of a cornucopia of marine organisms.[1] Penguins, seals, and whales plied the water, and seabirds filled the skies. Scientists hoisted large nets that teemed with a diverse population of invertebrates and fish, and finely meshed nets contained drifting sea organisms or plankton. These early discoveries paved the way for me and other scientists to study how marine invertebrates employ toxic chemicals to avoid predators.

I first encountered drifting sea organisms that sparked my interest in their chemical ecology in 1989 while I was a budding Antarctic scientist at McMurdo Station. Little did I know that a remarkable discovery lay in wait, one that would help launch my career. I was trained as a marine ecologist who studied organisms that live on the seafloor and was therefore accustomed to looking down rather than up. Nonetheless, my interests in how marine invertebrates defend themselves with chemistry took precedent over their location. One day, while visiting the dive hut our research team has established on the sea ice, I caught a glimpse of several half-inch orange sea butterflies—or *Clione antarctica*, winged snails—swimming about in the water in the ice hole nestled below the opening cut in the hut floor. Sea butterflies come in two varieties, shelled and unshelled, and both belong to a larger group of invertebrates known as mollusks, also represented by the more familiar clam, squid, and octopus. Lying prostrate on the floor of our dive hut, I peered into the clear frigid water and watched the butterflies swim below the ice using their delicate wings fashioned from flaps of mantle tissue to propel them. The shell-less butterflies were a conspicuous orange color, and despite their rapid wing beats they were no match for quicker predatory fish. However, their collective traits—the orange color, sluggish movements, and shell-less nature—in all likelihood translate as "I don't taste good!" Using a fine-mesh net, I extracted several of the butterflies and placed them in a plastic five-gallon bucket of ice-cold seawater. I then headed off at a brisk pace with my specimens to the Aquarium Building located near the transition between land and sea ice at the base of McMurdo Station.

Upon entering the Aquarium Building, I bumped into John Janssen, a biologist studying the behavior and physiology of Antarctic fish, and now a professor of biology at the University of Wisconsin–Milwaukee. John stood near a large circular seawater tank containing a school of "borks" (*Pagothenia borchgrevinki*), silvery, foot-long, trout-shaped Antarctic fish that feed on plankton. The majority of Antarctic fish live on the seafloor, but borks live in the water column, hiding from sea predators in the cracks and crevices of the sea ice. John joined me as I dipped my hand into the bucket to scoop up one of the sea butterflies and drop it into the tank of borks. One of the fish responded immediately, swimming up to the butterfly and engulfing it fully in its mouth. We didn't have to wait long to gauge a response. Almost immediately, the fish spat the butterfly out, turned, and swam off. The naked butterfly, now beating its wings, was none the worse for wear. A second bork approached the sea butterfly, sucked it into its mouth, and it too spat it out. A third and then fourth fish repeated the taste test with the same result. I removed the sea butterfly from the tank and dropped in a fresh butterfly from my bucket. John and I watched the pattern repeat itself once again. We tried it with a third butterfly—same thing.

Later that day while sorting plankton, John came across several amphipods—tiny shrimplike crustaceans whose name is derived from their two different (amphi) types of feet (pod)—with small orange balls on their backs. John removed one of the balls and watched it open its wings and swim away, revealing itself to be a sea butterfly. John showed me his discovery and mentioned that this particular species of amphipod was one of the borks' favorite foods. It was then that we had our "Eureka!" moment—we both realized that we could be on the brink of a truly amazing discovery. Were these amphipods carrying live sea butterflies for added protection because their predators wouldn't eat them? If so, we had uncovered nature's first example of one species abducting another species and carrying it around for chemical protection. There had truly never been any observations of this kind of behavior anywhere else. But before we could draw conclusions from our casual observations in the fish tank, we had much to do.

Even though John had found several amphipods with butterflies on their backs in his plankton samples, we needed to find out just how common this behavior was in nature. If carrying were common, it would suggest that butterflies were important to amphipods, adding some advantage to their survival. To test this theory, we designed a program to systematically sample amphipods in McMurdo Sound. We visited the fish and dive huts in various regions of McMurdo Sound and deployed cone-shaped plankton nets in the ice holes drilled beneath each hut. We suspended two-and-a-half-foot long plankton nets at both coastal and offshore sites and at depths that ranged from thirty to one hundred fifty feet. We attached each net to a swivel so that the prevailing current could flow through it, ballooning it out like a wind sock on an airport runway. We retrieved the nets twenty-four hours later and examined the contents, and we discovered hundreds of amphipods carrying sea butterflies. In nets from the coastal site, over 70 percent of the amphipods had sea butterflies attached to their backs.

We found that amphipods were picky about the size of the butterflies they selected, choosing only those that measured half of their own body length. The reason for this behavior became obvious when I happened upon an amphipod in an aquarium trying to carry a sea butterfly twice its size. The larger abductee seemed unperturbed by its smaller abductor, opening its wings and swimming while barely noticing the little amphipod hanging on for dear life. When, more often, an amphipod captured an appropriately sized sea butterfly, the amphipod would rotate the butterfly onto its back and then grasp it firmly with the pointed tips of its last two pairs of *pereopods* (opposing legs). The amphipods with their attached butterflies reminded me of children carrying their backpacks on the way to school.

Now confident that Antarctic amphipods carried sea butterflies, John and I conducted a series of carefully replicated experiments in the Aquarium Building at McMurdo Station to evaluate this unique behavior. First, we ran a series of feeding experiments and concluded that borks consistently spat out the butterflies. Later, we discovered that butterflies harbored a novel chemical that the fish find distasteful.[2] Then we ran another series of feeding experiments in which we observed that borks consistently ate amphipods when given the option. We then presented amphipods carrying sea butterflies on their backs to fish. In over 90 percent of these trials, the fish immediately spat out the amphipod carrying its sea butterfly, thus confirming our original hypothesis that amphipods carried sea butterflies for protection. Still coupled, the amphipods swam merrily away from their encounters with the borks, still carrying their winged guardian.

Clearly, amphipods got a great deal out of abducting and carrying a sea butterfly. But was there a cost to this behavior? John, a skilled videographer, designed a camera arrangement that allowed us to film individual amphipods swimming with or without sea butterflies attached to their backs. We discovered that amphipods with butterflies swam about 40 percent slower than those without. John and I surmised that because amphipods often prey upon mobile invertebrates, chasing down prey while carrying butterflies half their own size on their backs could be a hindrance. Imagine an Australian Aborigine hunter trying to run down a kangaroo while carrying a sixty-pound backpack. Nonetheless, given the preponderance of amphipods carrying butterflies, the advantages of packing a butterfly must outweigh this limitation.

But what of the fate of a sea butterfly that suddenly finds itself in the grasp of an amphipod? John and I couldn't come up with a single advantage for the butterfly in this unique relationship. But at least the sea butterflies' abduction appeared to have a happy ending: we never came across a butterfly that did not open its wings and swim away after we removed it from its amphipod. Despite our subsequent aquarium observations that amphipods carry sea butterflies for at least several weeks, it appears that the sea butterflies do not die in transit. Both our aquarium observations and field studies indicate amphipods periodically free their captives—carrying a healthy butterfly must provide a more potent chemical defense, or the two parties may have a mutual agreement: I will allow you to carry me for your own protection if you promise to set me free.

The Antarctic amphipod–sea butterfly relationship had tremendous appeal to the public at large. Our discovery of this relationship was a poignant reminder that marine scientists still have much to learn about behavioral and ecological interactions among plankton—in Antarctica and elsewhere. In addition to a publication of our paper in *Nature*,[3] our discovery resulted in enthusiastic phone interviews with reporters and journalists that generated articles in popular science media such as

*National Geographic* and *Discover* as well as many newspapers including the *Chicago Tribune* and *Los Angeles Times*.[4] In retrospect, what was
special about this eureka moment for John and me was that this discovery would never have been made if our two parallel backgrounds had not
intertwined at McMurdo Station at the same time. Indeed, the history of
scientific discovery is fraught with examples that owe their ontogeny to
the proximity of complementary minds coupled with a dash of seren-
dipity. The tale of the amphipod and the sea butterfly is but one small
link in a lengthy chain of interactions—some known and some yet unknown—that shape Antarctic planktonic communities. From the smallest bacteria (bacterioplankton), to single-celled plants (phytoplankton),
even larger invertebrates including copepod and amphipod crustaceans,
shrimplike krill (zooplankton), and salps—jellylike sea squirts about
the size and shape of a walnut—and jellyfish (gelatinous zooplankton),
the Antarctic plankton community functions in a number of important
ways. Some absorb and store greenhouse gases such as carbon dioxide;
others recycle nutrients. Many transport organic and inorganic carbon
to the deep sea in what marine biologists call a biological pump, which
moves dead organisms (plant and animal), feces, and other organic materials from the sunlit surface to the deep sea. As Antarctic waters are
rich in dissolved nutrients, plankton communities can be enormously
abundant in the presence of sufficient sunlight: if they are phytoplankton, they absorb the nutrients and photosynthesize; if they are zooplankton, they consume the phytoplankton.

Research teams at McMurdo Station strive to complete their diving
operations before the latter half of the austral summer when the phytoplankton, starved for daylight and living in low densities under snow-
covered sea ice during much of the year, erupt in a reproductive blossoming
known as a "plankton bloom." This photosynthetic frenzy turns the sea

a pea-soup green, and underwater visibility, which normally extends for hundreds of feet, plummets to less than five. Scuba diving in these conditions is challenging if not dangerous, and divers tether themselves with ropes so they won't lose their way. The explosion of phytoplankton in McMurdo Sound is followed by an eruption of phytoplankton-consuming zooplankton. Then, toward the end of the summer, as day-length diminishes, the plankton disappear as quickly as they arrive. Phytoplankton and zooplankton that aren't consumed by filter-feeders ranging from one-hundredth-inch copepods to one-hundred-foot blue whales drift to the seafloor to nourish hungry sponges, soft corals, feather stars, and sea squirts. Bacteria decompose and recycle the leftovers.

Ambient sunlight penetrating the waters 2,500 miles northwest of McMurdo Station near Palmer Station is sufficient to support marine plankton communities throughout much of the year because day-length along the Antarctic Peninsula does not vary to the extremes seen in McMurdo Sound. Each year, when the sun sets at McMurdo Station for the final time in the early winter, the station personnel do not see it rise above the horizon again for at least three long, dark months. In contrast, on the shortest day of the year, five hours of daylight occur at Palmer Station. As a result, plankton blooms are not as intense as they are in McMurdo Sound, but their persistence makes up for it. Given their dependable availability along the Antarctic Peninsula, plankton is essential in the diets of higher marine organisms. For example, seabirds, penguins, seals, and whales are dependent upon krill, a dominant zooplankton. Accordingly, because the Antarctic Peninsula is by far the most rapidly warming region of Antarctica and plankton is available to study throughout much of the year, no region of Antarctica is better suited to evaluate the first impacts of climate change on the ecology of plankton communities.

Dr. Hugh Ducklow—"Duck" to his friends—is a biological ocean-ographer and director of the Ecosystems Center at the Marine Biological Laboratory in Woods Hole, Massachusetts. His research expertise is centered on studies of marine bacterioplankton. Duck is also the current director of the NSF Antarctic Palmer Long Term Ecological Research Program (LTER), a multifaceted study initiated in 1990 that brings to-gether marine scientists with backgrounds in a wide variety of inter-woven fields ranging from oceanography to ecosystem modeling. The Palmer LTER focuses on understanding how the annual sea ice along the Antarctic Peninsula is linked to biological interactions across the food web. Duck explained to me that the project includes a strong em-phasis on plankton communities and encompasses the three primary groups—bacterioplankton, phytoplankton, and zooplankton. From the perspective of evaluating the impacts of climate change on Antarctic plankton, Duck feels that the timing of the Palmer LTER program could not have been better. Its twenty-year history coincides with a period of dramatic climate change along the Antarctic Peninsula that includes changes in temperature and in wind patterns, humidity, precipitation, snowfall, glacial melt water, and the extent of the annual sea ice. Each summer, a team of Palmer LTER scientists embarks from Punta Arenas, Chile, on a six-week survey cruise aboard the *Laurence Gould*. They travel to a series of twenty-five to thirty offshore sampling sites that span 560 miles of the western Antarctic Peninsula. Each year, resident LTER scientists at Palmer Station sample a second set of more localized coastal sites, including several penguin and seabird rookeries.

One of my current doctoral students, Julie Schram, participated in one of these LTER survey cruises and described to me the carefully or-chestrated series of activities undertaken at each sampling site. First, the participants deploy three-foot-long torpedo-shaped finned instruments called "spears" to measure levels of underwater light. As soon as the spear records the light measurements, the scientists decide upon a series of water depths in which to sample the spectrum of underwater light. Then they turn to a device that is the workhorse of oceanographers—the CTD, short for "conductivity," "temperature," and "depth"—to take measurements of the seawater. The CTD houses sensors that continuously transmit information to the technicians, including depth, pressure, temperature, conductivity (salinity), and density. Attached to the CTD are up to twenty-four three-foot-long plastic cylinders, each with a top and bottom cork that open when the winch operator lowers the CTD into the sea. At the desired depth, technicians trigger the corks to close, trapping a sample of seawater within each. Back on ship, technicians remove the samples of seawater from the cylinders to evaluate and measure various characteristics of the microbial bacterioplankton and phytoplankton. Duck and his team employ instruments used for biomedical technology—flow cytometry and DNA sequencing to enumerate and identify the bacteria recovered from seawater samples. To determine the quantity of phytoplankton in a given collection of seawater, the technicians measure fluorescence, a bio-optical property of chlorophyll pigments that provides an indirect measure of the amount of phytoplankton.

While these often-used techniques of sampling seawater and evaluating its proportions of bacterioplankton and phytoplankton are providing key information for Duck and his team of LTER scientists and technicians, newer technologies are opening up exciting paths to studying ecosystem processes along the western Antarctic Peninsula. Beginning in 2007, the Palmer LTER program has employed unmanned advanced automated underwater vehicles to expand their studies of phytoplankton along the Antarctic Peninsula. Three so-called "gliders" now exist in the Palmer LTER fleet, two for diving to depths up to three hundred feet and one for diving much deeper, up to three thousand feet. Each glider is about five feet long and looks like a bright-yellow jet plane with a torpedo-shaped fuselage, two fins resembling wings, and a tail equipped with a vertical fin and rudder. The gliders carry a scientific payload of instruments that can measure salinity, temperature, oxygen, and fluorescence.

The gliders are efficient in their ability to operate as they don't require a fuel-guzzling motor or a propeller that can become tangled in seaweed. Instead, they zigzag up and down the water-column using their fins to translate changes in buoyancy into horizontal motion. To do this, submersible engineers cleverly designed a device, powered by a small battery, that forces mineral oil in and out of an inflatable bladder. To control pitch, the degree to which the vehicle points upward or downward in the water, the battery can slide forward or backward, shifting the center of gravity of the glider's fuselage. To control roll, the degree to which the vehicle leans to the right or left, the battery can slide left or right. In 2009, a glider similar to those used in the LTER program made history when it became the first robotic device to cross the Atlantic Ocean. College students from Rutgers University remotely navigated the glider to trace backward the path of Christopher Columbus's ship, the *Pinta*, between North America and Spain. The glider surfaced three times a day to check its location, download data, and receive piloting instructions via a satellite phone on its tail. The crossing took 221 days and spanned an impressive distance of 4,600 miles.[5] Even the prestigious exhibition of the glider at the Smithsonian Museum in Washington, D.C., doesn't do justice to the significance of this historic event. The execution of the Rutgers glider signifies a breakthrough in remote underwater vehicle technology that is allowing scientists for the first time to take measurements of key oceanographic features in real time. This technology is transforming the way humankind studies the seas.

Two of the program's three gliders are kept at Palmer Station, where scientists engage in studies related to understanding patterns of phytoplankton production. Scientists use one glider to survey and collect information above a submarine canyon near the station. The LTER scientists think that this submarine canyon and others like it are responsible for connecting deep, nutrient-rich seawater to surface waters that fuel rich Antarctic phytoplankton blooms. Rather than collecting phytoplankton from a single location at a single time aboard a ship, scientists using the glider are able to study the submarine canyon's seawater in many locations. The treasure trove of data generated by the glider will help the Antarctic scientists understand what triggers plankton blooms and how the resultant phytoplankton provide nutrients and energy to zooplankton and larger marine animals. The second, deep-diving glider is deployed in the waters between Palmer Station and the British Antarctic Survey's Rothera Station—a distance of 224 miles—to measure and transmit physical and bio-optical oceanographic information. Duck explained to me that the third glider travels aboard the *Gould* on survey cruises. Like a hunting dog on point, the glider is released a few days ahead of the ship so it can transmit data back that helps Duck and the team of LTER scientists decide their route and which station to sample next within the LTER grid that includes about fifty stations. As the glider swims along at one knot, eventually the ship catches up and the crew recovers the glider. As it takes a day or two for the ship's science crew to complete the sampling at a given station, the glider may then be redeployed to hunt once again during this time.

Duck explains to me that the Palmer LTER is among the first large-scale studies investigating the role of marine bacteria in a polar sea. Bacteria are smaller than most cells but slightly larger than viruses. They also happen to be very abundant at sea—a teaspoon of seawater contains more than a million bacteria. Technicians on the *Gould* collect seawater from the CTDs to measure rates of bacterial metabolic activity. To do so, they inoculate the bacteria with radioactive tracers and then measure the amount of radioactivity the bacteria absorb over a few hours. After studying this, they can calculate how fast the bacteria make proteins and divide to form new bacterial cells. With this data, the scientists can answer big questions about how marine bacterial activity varies from the sea-ice-free northern waters to the ice-covered southern waters of the Antarctic Peninsula, and from near-shore waters freshened by glacial meltwater to offshore seawater. Furthermore, microbiologists can also address what besides temperature affects bacterial activity and how important this activity is in recycling organic matter.

Polar seas experience vast blooms of the tiny plant cells that make up phytoplankton, blooms as rich in organic plant material as those that occur in the estuaries of the Chesapeake Bay or the Gulf of Mexico. Where does all this plant matter go? Duck explains that zooplankton along the Antarctic Peninsula consume much of this plant matter, and they in turn feed larger animals. Some of the plant matter also sinks to the seafloor where it serves as long-term storage for atmospheric carbon dioxide. But in Antarctica, the amount of plant matter that succumbs to bacterial decomposition is actually quite small compared to that in bays and estuaries and the Gulf of Mexico: only about 10 percent of what is produced. This number differs dramatically from the 50 to 75 percent bacterial decomposition rates of plant matter in warmer seas. Duck believes that polar seas are fundamentally different in this respect than warmer seas, suggesting that polar seas have naturally evolved to deliver the majority of plant matter to hungry zooplankton, which in turn nourish fish, penguins, seals, and baleen whales. Duck and his team of microbiologists are hoping to answer the critical question of whether rapid climate warming is altering this fine-tuned polar food chain.

Duck's first winter cruise along the Antarctic Peninsula as the director of the LTER program turned out to be a challenge. The *Nathaniel*

*Palmer* departed from Punta Arenas, Chile, in early September, and would stop at stations so that scientists could take samples when the sea ice was at its maximum extent. In a strange series of sad events, the day the ship's occupants first sighted Antarctica was September 11, 2001, and the ship's crew and scientists had to settle for fragments of information about the terrorist attacks. Several days later, soon after the ship departed from a stop at the British station Rothera, the biology laboratory there burned to the ground. And if these events weren't enough, in early October the *Palmer* was beset in heavy sea ice in Marguerite Bay along the central western Antarctic Peninsula. Much like the early-twentieth-century Antarctic explorers of the heroic era, those aboard had no idea when their ship might be released from the grasp of the ice. All research activities were curtailed as they drifted helplessly for an entire month. As one of the lead chief scientists, Duck, along with several of his colleagues, had to invent ways to keep everyone occupied and in good spirits. Despite having plenty of food and enough fuel to provide electricity, Duck had to exhibit critical leadership skills such as those exemplified by polar explorer Ernest Shackleton when he and his twenty-seven men were beset for months on the frozen Weddell Sea in 1914. Finally, in early November 2001, one of the *Palmer*'s crew detected a nearby lead in the sea ice from a satellite image (a tool that Shackleton would have found eminently useful). With this information in hand, the captain skillfully maneuvered through several ice ridges into the lead and, to everyone's delight, eventual freedom.

Once the team of science technicians on the Palmer LTER survey cruise have finished their collections of seawater for bacterioplankton and phytoplankton, they move up the food chain to zooplankton. Debbie Steinberg, a professor at the Virginia Institute of Marine Sciences and a former research colleague of mine at the Bermuda Biological Laboratory, is currently the resident zooplankton expert on the Palmer LTER program. She explained how the zooplankton was collected and studied onboard the *Laurence M. Gould*. Two square-shaped nets deployed to a depth of 390 feet are towed by the ship at one knot for a period of about thirty minutes. The smaller of the two nets measures two-and-a-half by two-and-a-half feet, and captures tiny zooplankton like copepods in its fine mesh. The larger net measures five feet by five feet and has a slightly larger mesh size that collects bigger zooplankton like krill, salps, sea butterflies, and amphipods. Despite performing dozens of net tows on a given cruise, Debbie describes each net hoisted out of the water as "magical." Like children opening a Christmas present, or a young naturalist flipping over a rock in search of a colorful insect, the eager scientists experience a renewed sense of anticipation. What will be in the net this time? Will it be salps again? Are the sea butterflies shelled or naked? Could there be something unusual, even weird, in the net? One day, Debbie and her team discovered a never-before seen fish— white with a black cape-shaped patch on its torso. Debbie subsequently nicknamed it the "Dracula Fish."

The overarching question regarding zooplankton that intrigues Debbie and the other scientists working on the Palmer LTER project is this: how have the zooplankton communities along the Antarctic Peninsula changed and how will they continue to change? And as a corollary, if there are changes, how might this affect how nutrients are recycled in this region, and how might it affect the predators of zooplankton, such as seabirds, penguins, seals, and whales? Krill can form immense schools that scientists measure in hundreds of square miles. Early Antarctic explorers wrote about seas so rich in krill that they turned red. Individual krill trigger this unique schooling behavior by producing a bluish bioluminescence that serves as a visual attractant to other members of the same species—like fireflies on a suburban lawn on a warm summer evening.

Once the nets loaded with zooplankton are retrieved from the sea, Debbie and her team sort their collections and count their specimens. While they mostly catch salps and krill, they also retrieve copepods, arrow worms, jellyfish, and a group of amphipods that live on the bells of jellyfish and other gelatinous zooplankton. The word *copepod* is derived from the Greek term *kope*, and translates as "oar feet," a misnomer for a teardrop-shaped crustacean about as big as the head of a pin that actually uses one of its two pairs of antennae rather than its feet to "row" itself along in a series of jerky motions. The reason for this stop-and-go movement is that copepods are so diminutive that the water in which they float is viscous to them, as if they were swimming in a liquid the consistency of Jell-O. Copepods live everywhere—in both saltwater and fresh water; on wet forest floors; in swamps, water-filled cups of bromeliads and pitcher plants; and in all the world's oceans. Copepods that live among plankton are so abundant that some marine biologists believe they comprise the largest collective biomass of any animal on the planet, although others argue that Antarctic krill have this distinction. Myrmecologist E. O. Wilson might counter that the world's entirety of ants would qualify for the title. Regardless, the ubiquity of copepods makes them a key player in the food webs of the Southern Ocean as they nourish small fish, seabirds, krill, and baleen whales, and their feces and decomposing bodies transport carbon to the deep sea. I am quite familiar with the copepods that live along the coast of Alabama as each fall I take my invertebrate zoology students on a weekend field trip to Dauphin Island Sea Laboratory on the Gulf of Mexico. A trip aboard the lab's eighty-foot research vessel *Alabama Discovery* gives the students the opportunity to experience a plankton tow. After the ship's mate empties the contents of the plankton net's cod end into a glass jar, the students take turns holding the jar up against the sun to better see the copepod soup—experiential learning at its best.

Antarctic copepods need to be vigilant. Arrow worms, a group of highly predatory marine invertebrates, stalk these plankton. Ranging from one-tenth of an inch to four-and-a-half inches in length, the taxonomic classification of these transparent, dart-shaped animals has puzzled scientists for years. Arrow worms don't quite fit on either side of the two major branches of the tree of life. Are they more closely akin to the branch with the nematodes (small wormlike animals) with whom they share some anatomical characteristics or the other branch—where they share embryological traits? They are certainly not "worms" in the sense of earthworms—and despite having an outer skin (cuticle) that they periodically shed, they do not belong to the crustacean family. Recent DNA studies suggest arrow worms share a closer affinity with nematodes (roundworms). These ferocious predators of zooplankton undulate their bodies like competitive swimmers doing the butterfly stroke. They chase down their prey, grasp them with a set of hooked spines, and immobilize them with a swift injection of tetrodotoxin. One milligram of this potent neurotoxin can kill a human. This same toxin exists in a diverse variety of organisms, including marine snails, the eggs of horseshoe crabs, frogs and salamanders, the blue-ringed octopus, and the skin and organs of the puffer fish.

Antarctic seas are home to an impressive sixty-seven species of hyperiid amphipod—quarter-inch crustaceans that hitchhike on gelatinous zooplankton like jellyfish, salps, comb jellies, and sea butterflies. "Hitchhike" may be too benign a term. Some hyperiid amphipods parasitize the tissues of their hosts; others are parasitoids, laying their eggs in the jellies so that their young can consume the host when they hatch. Still others, perhaps the true hitchhikers, appear to use jellies simply as drifting platforms upon which to live, but prefer to feed on phytoplankton, copepods, krill, fish eggs, and larvae. Variations on the hyperiid amphipod/jelly theme include *Hyperiella dilatata*—the species of amphipod that John Janssen and I discovered carrying sea butterflies for protection. Zooplankton experts have long wondered about the evolution of this unique hitchhiking of amphipods on jellies. Besides the obvious benefit for those amphipods that make a meal of the jellies upon which they live, jellies might also provide amphipods protection from fish predators, either by giving them a place to hide from fish or because fish avoid the stinging tentacles. Riding along on a jelly also provides access to a smorgasbord of smaller plankton that drift by and can be captured and consumed, and the energy-saving ride sure beats exhaustive swimming.

One of the most common hyperiid amphipods in the Southern Ocean is *Themisto gaudichaudi*. Like all hyperiids, *Themisto* hitchhikes on jellies and has huge eyes to help it locate its prey. Near King George Island to the north of the Antarctic Peninsula, marine scientists have found huge swarms of *Themisto* dining on phytoplankton, copepods, small krill, and sea butterflies. The high energy content of *Themisto* makes it an attractive prey, and it is among the main food resources for a variety of Antarctic fish, squid, birds, and whales. As such, *Themisto* provides an important ecological link in the food web between small zooplankton and larger consumers.

Debbie and her team preserve most of the zooplankton in alcohol and store them in labeled plastic bottles, but they keep some of the salps and krill alive in holding tanks so they can measure fecal pellet production, or what Debbie amusingly refers to as "poop experiments." Salps or krill are placed individually in containers of seawater and kept in a walk-in cold room where they're fed a diet of phytoplankton for a period of ten hours. Their droppings are then collected for analysis. The team measures the nutrient and energy content of the feces and compares it with that of the phytoplankton they consume, providing important information about how efficiently the salps and krill digest and utilize their food.

When Langdon Quetin and Robin Ross, professors from the University of California–Santa Barbara, were leading the zooplankton team for the Palmer LTER program, they carried out a number of "instantaneous growth measurements" on krill. Technicians placed krill of varying body sizes in individual containers of seawater for four days, checking the containers every six hours. As Antarctic krill shed their exoskeletons every nine to twenty-eight days in a process called *molting*, the odds are pretty good that a certain portion of the krill will molt during this time period. By comparing the size of the exoskeleton with the size of the animal from which it came, scientists can determine if an individual krill is in a shrinking or growing phase. Remarkably, Antarctic krill are one of the few animals capable of actually shrinking. For example, if the molted exoskeleton is bigger than the krill that produced it, then that individual had clearly shrunk, and vice versa. By carrying out these experiments with various-sized krill in different seasons over a period of several years, Langdon and Robin were able to determine their growth patterns. One remarkable outcome of the experiments was the discovery that Antarctic krill shrink during the winter months when they adopt a "starvation mode," during which they cease to feed and rely instead on stored energy reserves. In a sense, Antarctic krill, like Arctic polar bears, hibernate.

Additional Palmer LTER zooplankton experiments examine the rate at which salp or krill feces sink. Not a particularly hot topic around most dinner tables, fecal-sinking rates are nonetheless important because, with billions of zooplankton living in the water column, their feces represent an important contribution of nutrients and energy to seafloor communities. By measuring how fast feces sink and estimating their rate of decomposition, biological oceanographers can determine how much the feces of a given zooplankton species contribute to the energy budget of the deep sea. Recently, krill biologists were surprised to discover that more than krill feces find their way to the deep seafloor. Like tiny missiles, adult krill dive headfirst into the seafloor, stirring up organic particles and tiny organisms in the sediment that are then suspended above the ground. Krill then capture these food items in their filter-feeding baskets woven by appendages near their mouths. Marine scientists have long known that krill can move both horizontally and vertically; however, krill biologists had not considered that these water-column filter-feeders could descend thousands of feet to forage on the deep seafloor. This observation is a wonderful example of how discovery can trump conventional knowledge.

Marc Slattery, a former doctoral student of mine, made another unconventional discovery of an Antarctic plankton feeder. Now a professor of marine chemical ecology at the University of Mississippi, Marc studied the ecology of Antarctic soft corals in McMurdo Sound for his PhD research in the late 1980s. Knowing that soft corals are normally constrained to feed on plankton and that plankton in the Sound is in scarce supply most of the year, Marc positioned a remote underwater time-lapse video camera on a tripod so he could film a three-foot tall treelike soft coral (*Gersemia antarctica*) perched on the seafloor of New Harbor. When Marc dove under the sea ice and retrieved the camera a week later, the video told a remarkable story. The coral tree had not stood still with its branches extended to capture plankton as one would expect of soft corals. Instead, it had lain down and proceeded to slowly roll 360 degrees around the base of its trunk. As the branches of the coral tree brushed against the sediments, hundreds of tiny feeding polyps scooped up organic particles and tiny algae and invertebrates living on the surface. Once its rotation was complete, the coral tree stood back up. In yet another twist on how we perceive soft-coral biology, the entire coral tree then pulled its bulbous base from the sediments and crawled a distance of five to ten feet, replanted itself, and repeated its novel feeding behavior. I recall first viewing Marc's video of this unbelievable behavior—I might just as well have watched the oak tree in my front yard pull its roots from the soil and crawl next door. The march of the soft coral reminded me of J. R. R. Tolkien's Ents, those immortal walking trees that occupied the forests of Middle Earth.

In addition to taking biological, chemical, and physical measurements at sea, the Palmer LTER scientists evaluate satellite photographs that reveal the seasonal extent of the annual sea ice along the Antarctic Peninsula, along with information on air temperatures, wind directions and speeds, and humidity. In a 2009 article in the journal *Science*,[6] Martin Montes-Hugo and his colleagues describe a rapidly changing ecosystem for phytoplankton and zooplankton and the plethora of large marine animals along the peninsula that consume them. In their study, the scientists divided the seas along the western peninsula into northern and southern regions. In the northern region, where climate change has been most pronounced over the past thirty years, the environment has shifted from a polar climate to a moist, warm, subantarctic climate. The days have grown cloudier, the winds stronger, and elevated temperatures have greatly reduced the annual sea ice. Cloudier skies mean that less light filters through to support photosynthesis, and high winds turn over and mix the sea, pushing phytoplankton deeper, away from what little light still exists. As a result, less phytoplankton remain along the northern region of the peninsula.

Scientists have found that the nature of the phytoplankton has also changed. Larger phytoplankton such as diatoms (single-celled algae with silica shells) are being replaced by smaller species less favored by krill. With less phytoplankton, and of poorer quality, populations of krill along the northern region of the peninsula are disappearing. And unfortunately for the seabirds, penguins, seals, and whales, the krill are being replaced in this region by vast numbers of salps. Krill prefer cold waters with high densities of phytoplankton, and they depend on the presence of sea ice because, as juveniles, they feed on diatoms that grow on the undersurface of the ice. Salps, on the other hand, prefer slightly warmer seas free of ice, and despite being voracious filter feeders they ironically prefer regions where phytoplankton do not occur in high abundance. This regional preference is tied to how salps feed—they capture plankton on a sticky, mucous-laden filter-basket. As they swim using muscular pulsations of their barrel-shaped bodies, seawater is forced through their mucous filter. When phytoplankton becomes too dense, it clogs up the filter, making it difficult or impossible for the salps to process their food. With warmer seas, less sea ice, and fewer phytoplankton along the northern region of the peninsula, salps are thriving. The seabirds, penguins, seals, and whales that ply these seas find salps no substitute for krill. Akin to replacing steak with lettuce, salps, nutritionally, are essentially bags of water. Evidence of this nutritional stress on larger animals is mounting as krill populations are replaced by salps in planktonic ecosystems. In a 2011 article, Wayne Trivelpiece and his colleagues attribute plummeting penguin populations, which have decreased by more than half since the 1980s at their study sites on King George Island, to the 80 percent decline in krill in the Scotia Sea off the northern tip of the Antarctic Peninsula.[7] And Oscar Schofield, one of the Palmer LTER principal investigators, pins the dramatic decline in krill squarely on the rapid climate-change-induced depletion of phytoplankton, the crucial lower rung of the food chain.[8]

The Palmer LTER scientists report that the southern region of the western Antarctic Peninsula has also undergone environmental changes over the past thirty years, but not yet as dramatic as the northern sector. Cloud cover and winds have diminished, and as such more light is available for phytoplankton. Krill are also doing well. But with warming temperatures, the ongoing recession of the annual sea ice along the southern region of the peninsula may begin to put pressure on populations of juvenile krill, given their dependence on the gardens of algal cells that thrive on its undersurface. Duck suspects that by mid-century the pace of climate change will spread to the south, challenging zooplankton and, by default, the large iconic marine animals that humankind has come to synonymize with Antarctica.

# Chapter 4

*Antarctic Seafloor*

## AN OASIS IN THE DESERT

*A*cross the seafloor, peach-colored soft corals spread out amid cream-tinted sponges standing three to four feet tall like ancient Venetian vases. Giant marine worms and twelve-inch-diameter sea spiders added an element of absurdity to the landscape. Bright red sea urchins carpeted the seafloor, and red and yellow starfish nestled among them, feeding on sponges. Among the soft sediments resided fields of clams and snails, while shelled brachiopods and bushy colonies of bryozoans attached to the exposed hard surfaces. Tiny orange sea butterflies swam about, flapping their wings, and a Weddell seal, attracted, perhaps, to the breathing hole beneath our dive hut, approached.

Instead of the luxuriant kelp beds off the coast of California or the tropical coral reefs of Caribbean islands where I had become an experienced diver, I was floating below a six-foot layer of annual sea ice, surrounded by the bone-chilling waters and rich diversity of sea life of McMurdo Sound. In the midst of such awesome beauty, I could have lost all track of time but for my painfully throbbing fingertips and toes reminding me that my thirty minutes under the ice were just about over. This was my first ice dive in Antarctica in 1985. My location, Ross Island, 2,300 miles due south of New Zealand, was as close to the South Pole as a human being can reach and still dive under the sea ice.

My dive partner and the dive officer, Rob Robbins, and I had spent a lengthy two-and-a-half hours in the dive locker, checking over our twin seventy-two-cubic-foot scuba tanks, U.S. Divers Royal Aquamaster two-hose regulators, and weight belts and ankle weights, and donning the DUI or Viking dry suits that would presumably protect us from the frigid 29 degree Fahrenheit Antarctic waters. My heart raced and my palms sweated as our Spryte, a common tracked vehicle, crossed the land/sea ice transition in front of McMurdo Station, pulling up to a dive hut only a few hundred feet offshore but perched over one hundred feet of water. I peered overhead, trying to see the holes that had been cut both through the floor of the dive hut and the sea ice, just large enough to accommodate our entry into the sea. A safety rope sporting colorful flags and a strobe light to ensure divers can find their way back to the ice hole dangled to the seafloor. Estimates of visibility in Antarctic waters range upward of five hundred to a thousand feet, an order of magnitude greater than those recorded even in tropical seas. I could see forever.

Fully suited, masked, and finned, with my breathing regulator safely secured in my mouth, I slipped into the water. I reached up with my left hand and compressed the valve on my right shoulder and released air trapped in my dry suit. The ice-cold water numbed my exposed face, but for now my hands, feet, and torso were comfortable. Descending first through six feet of sea ice and then, once below the ice, to a depth of about twenty feet, I paused to take in my surroundings. The sea ice above me glowed, filtering sunlight to the depths. I was drifting as if just below the ceiling of a magnificent building whose floor lay eighty feet below me. I was struck by the same sense of awe one experiences entering the Sistine Chapel, only instead of Michelangelo's paintings, I was gazing at a ceiling aglow and adorned with intricate platelets of ice.

Diving underwater in Antarctica, in a land known for its ice and miles of snow, is an important ingredient to carrying out marine science in this faraway land. Each field season, our research team would assemble in the dive locker where Rob Robbins would brief us, bringing us up to date on dive protocols, pointing out where we should store our gear, and assigning us a Spryte tracked vehicle equipped with special racks to secure our scuba tanks. If any new divers were on the team,

Rob would usually do a "check-out dive" to ensure that despite their extensive pre-Antarctica dive training they were proficient under the ice. While the check-out dives were designed to ensure divers were comfortable in a dry suit and able to follow ice diving safety protocols, Rob also employed psychological measures to evaluate a diver's comfort level. If a diver appeared confused, distracted, or nervous, Rob might choose to schedule an additional check-out dive. Under-ice diving, just like cave-diving, requires composure under sometimes challenging conditions.

Our goal was to collect marine invertebrates, and we were eager to launch into our research. To expedite our collection process we established one or two dive huts in front of the station, which were positioned directly over thirty-six-inch-diameter holes drilled through the six- to eight-foot-thick sea ice. The huts allowed our dive team to make repeated collections of marine organisms without having to drill a hole through the ice every time we needed to dive. The wooden huts were about fifteen feet long and twelve feet wide and were painted bright orange to make them visible in blowing snow, a weather condition we soon learned to refer to as a "white out." Each hut harbored a bench, a countertop for a propane camp stove, and a few shelves to store granola, energy bars, hot chocolate, and other provisions. The underbelly of the huts sported tow bars and pairs of skids to allow them to be pulled across the sea ice.

Each hut had a four-by-four hole cut in the middle of the wooden floor, and in the corner, a small Preway heater that resembled a pot-bellied stove with an exhaust pipe that exited through the ceiling. The heater burned diesel fuel and ran constantly to keep the hut warm for visiting dive teams and to prevent the ice hole from refreezing. One year the wind blew so hard that a heater in one of our dive huts back-drafted, forcing hot air down the exhaust pipe and back through the heater. The unoccupied hut caught fire, but by the time the McMurdo

Fire Department arrived it was completely engulfed in flames, and the fire squad couldn't do much but watch it burn down to a few remaining coals and blackened metal beams. The burning dive hut reminded us that fires are a constant threat to scientists and support staff, as the dry, cold, and windy conditions could easily promote an uncontrolled blaze. Architects design polar stations with the structures sufficiently separated to ensure that if one building burns down, there will at least be an untouched building where personnel will find shelter and electricity until help can arrive. In November 2003, the biology laboratory at the British Rothera station—valued at over two million British pounds—burned to the ground when an electrical short circuit sparked a flame. The winds were so high during this fire that attempts to extinguish it were fruitless. While the biology laboratory was eventually rebuilt, the fire set back important climate change research and other biology projects.

We had a pretty good idea where the various marine invertebrates near the station were located, both because of our anecdotal observations from past years and surveys by other dive teams. As such, we could position our huts to best meet our collection needs. But establishing our dive huts was no trivial matter. We first had to select a hut from the storage lot. The drill operator then had to haul the hut out on to the ice—a box of Snickers bars was usually a sufficient bribe. The drill rig was an impressive contraption that looked like a huge tractor with a large, hydraulically operated drill secured to its back end. Once on site, the rig operator would unhook the dive hut and climb up into a separate elevated seat where he manipulated the various switches and levers for the drill. With the rig's engines revving loudly and the three-foot-diameter bit spinning madly, the drill would penetrate into the sea ice. After about fifteen minutes, seawater would gush to the surface of the sea ice surrounding the hole. With a final pumping motion of the drill, the operator would clear away the last of the loose ice. We'd then drop a weighted rope to the seafloor to ensure that the water depth was between eighty and one hundred feet. After clearing away the piles of ice around the hole, the operator could position the hut so that the hole in the ice was directly below the square opening in the floor of the hut. The final position of the hut depended on the direction of the prevailing wind. The high winds demand that care must be taken when entering and leaving dive huts. Doors blown open or slammed shut by one of those gusts can break human limbs. To ensure a backup in case of a dive emergency, the operator drills a second "safety hole" about ten meters from the dive hut and we fit the opening with a Styrofoam lid to prevent refreezing.

We inevitably needed to travel farther from the station to expand our collections. For several years in the late 1980s, the National Science Foundation experimented with deploying science teams into the field using a hovercraft. Riding on a cushion of air generated by a large propeller, the craft, essentially the size of a large extended pickup truck, flies across the smooth surface of the sea ice at thirty to forty miles per hour. It turns out that the hovercraft, which is operated by two professional pilots, was a marvelous way to get our dive team and all their gear out onto the field. Divers can load up their dive tender, dive gear, and several coolers for collected samples and head out across the sea ice in search of cracks large enough to accommodate diving. Usually, we could find these ice cracks along the boundaries of the shoreline or near small islands, like Hutton Cliffs and Little Razorback Island. In due course, we learned that seeing groups of Weddell seals sprawled on the sea ice was a good sign that an ice crack through which we could dive was nearby. The crack had to be big enough for the seals to haul their thousand-pound bodies to the surface. After locating a suitable crack, we would don our dive gear within the heated confines of the hovercraft, allowing us to enter the icy sea with a warm body core, a huge advantage when diving outdoors. Unfortunately, since we were the only ones who used the hovercraft, the NSF eventually shipped it back to the United States where it would be of more use. We were sad to see it go.

Helicopters also played an important role in our field-based diving program. In the 1980s, the Navy maintained a small fleet of Huey helicopters that serviced McMurdo Station. They were used in large part to transport glaciologists, geologists, and terrestrial field biologists to distant field camps. Camping gear, food, and scientific supplies wrapped in a mesh net can be lifted off the ground and carried below the belly of the helicopter in what pilots call a "sling load." And most importantly, emergency medical evacuations of personnel from remote field camps are carried out by helicopter. On short day trips, pilots would fly us out to the leading edge of the sea ice. We always considered a visit to the ice edge a treat, as we could witness at any given time killer whales, leopard seals, Weddell seals, and emperor penguins. One year, I even dove with emperor penguins, an unforgettable experience. I watched them jump off the ice edge above me, and I recall that, underwater, they did not so much swim, as fly, past me. Streams of tiny bubbles from their feathers made them appear as if they were jet-propelled black-and-white torpedos.

Each year, our field team would helicopter across McMurdo Sound and spend a week at New Harbor, a scenic bay framed by McClintock Point to the north, Baker Point to the south, and the massive Taylor Dry Valley to the east. Sometimes, we camped in Scott tents, pyramid-shaped polar tents with double skins, but more often we occupied an annually established field camp constructed of several canvas Jamesway huts, sturdy dwellings that provided both separate diving, sleeping, and eating quarters. The helicopter pilot could not sling the heavy drilling rig across the Sound (it weighed more than the helicopter) and steep ice ridges rendered the fifty-mile drive across the Sound impassable.

Therefore, we had to utilize a different method to establish our dive holes: high explosives.

Quentin Rhoton was equally comfortable with a bottle of Jack Daniels or a stick of dynamite in his hand. Fortunately, he never combined the two. Trained as a demolition expert, Quentin looked like a Hell's Angel with his long, braided ponytail, beard, and generous mustache, but he had the happy-go-lucky nature of a puppy. He was the go-to guy at McMurdo Station when it came to working with high explosives, and he loved it when we invited him to New Harbor for the week, which got him "out of town," he would boast. Transporting dynamite from McMurdo Station to New Harbor required its own separate helo flight, containing no passengers, additional baggage, or, especially, extra fuel or detonators. I wondered if the pilots drew straws for those flights.

Using a Jiffy Drill, a handheld device that looks like an oversized gasoline-powered Weed Eater, we would take turns drilling the eight-inch-diameter hole through six to eight feet of sea ice. The drill bit came in two-foot sections, and as we drilled down, we would add additional sections. Once we had finished drilling, Quentin would slide a long bamboo pole into the hole, its lower half wrapped with some thirty sticks of prewired dynamite. Uncoiling a large spool of detonation wire behind us, we would retreat about three hundred feet from the blast site. At this distance, Quentin could make the final electrical connections to the dynamite detonator box. When everything was set, an eager graduate student would gleefully depress the box's plunger, triggering a deafening explosion that shot chunks of sea ice the size of baseballs, softballs, and even basketballs hundreds of feet into the air. What the students did not realize, of course, was what the hole in the sea ice would look like when we returned to the detonation site. Sometimes measuring as much as eight feet across, the ragged hole would be chock-full of heavy chunks of floating sea ice. Only after several hours of back-breaking, hand-chilling hard labor removing the ice with tongs, breaker bars, and shovels would the hole become sufficiently free to evaluate whether it would accommodate a diver. On occasion, the hole was too narrow and we had to dynamite again.

Overall, ice diving in McMurdo Sound has an enviable safety record since the first dive there on January 5, 1961, when photographer Jim Thorpe and engineer Don Johnson plunged into the water in dry suits, two pairs of thermal underwear, two sets of wool socks, wool gloves fastened with metal wrist clamps, and rubber hoods secured with metal O-rings.[1] Even taking into account the small number of divers, the incidence of accidents that occur in Antarctica is rare, in large part due to the rigorous training required of Antarctic research divers. Nonetheless, a few incidents and accidents have occurred over the years, some with happier endings than others. Once, two graduate student divers couldn't relocate the three-foot-diameter hole in the sea ice below their dive hut. One might surmise that in their enthusiasm to explore a new world they simply lost their bearings. Anticipating running out of air, one of the divers found a narrow crack in the sea ice near the shore where he had just enough room to push his head in and breathe. His partner wisely abandoned him at this point and continued to search for the dive hole beneath the dive hut, which he eventually found. This time, keeping careful track of his bearings, he returned to the crack and tugged on his partner's fins to let him know he had located the hole. Unfortunately, the student at the crack thought the tugs meant that his partner had run out of air and that he was attempting to pull him away from the crack so that he could breathe through it. Panicking, he refused to budge. In frustration, his dive partner swam back to the ice hole and yelled at the dive tender in the hut to run over to the shore and tell the other student that he had found the hole. Thank goodness he finally listened to reason, and both students lived to see another day. However, when their professor found out what had happened, he put them on the next flight departing McMurdo Station. They had learned a hard lesson—just as you should never let down your guard when walking or driving upon sea ice, never let your guard down while diving under sea ice.

And yet that lesson was better than the fate of Jeffrey Rude. Rude was an outstanding graduate student studying under Paul Dayton, the father of Antarctic marine ecology and a professor of oceanography at Scripps Institute of Oceanography in San Diego. Dayton's studies at McMurdo Sound in the late 1960s and 1970s had revealed a surprisingly rich and complex seafloor community where starfish and sponges were key players. Starfish consumed the most rapidly growing sponge type and thereby kept that sponge from crowding out other species. Populations of the sponge-eating starfish were themselves kept in check by other starfish that consumed their water-born larvae, holding their arms up into the water column and catching larvae with thousands of outstretched tube-feet. Some starfish also consume the sponge-eating starfish adults. Like a pack of hungry wolves, groups of ten to twenty individuals can attack and devour adults ten times their own size. The sponge-eating starfish occupy a group considered a keystone species. If one were to remove the keystone starfish from the food chain, the fastest growing sponges would take over the seafloor, crowding out other species and fundamentally altering the entire community.

On October 12, 1975, when Jeffrey was in the early stages of his doctoral studies on the reproduction, feeding tactics, and mechanisms of energy storage of Antarctic marine sponges, tragedy struck. He was traveling across the sea ice near McMurdo Station when the tracked vehicle in which he was riding partially broke through the sea ice. The vehicle's occupants jumped safely out onto the surrounding ice, but Jeffrey, realizing he had forgotten his camera, climbed back inside the vehicle to retrieve it. At that very instant the sea ice broke, and the tracked vehicle immediately sank hundreds of feet to the seafloor. Jeffrey never had a chance. Thirty-six years later, Paul Dayton still laments the loss of his student: "At the time of his death [Jeffrey] had found a deep interest in sponges and a passion to pursue a research career with such a still poorly understood group of organisms. . . . As with all premature deaths of bright young people, one can only wonder what might have been." Had Jeffrey lived to publish his work, it would have provided a foundation for our later studies on the chemical defenses of Antarctic sponges. For instance, Jeffrey's goal to measure the sponge's energy-storing capacity would have allowed us to more quickly and efficiently evaluate whether sponges that invest in costly chemical defenses are those with the greatest energy resources. A plaque in Jeff's honor is now displayed in the Albert P. Crary Science and Engineering Center at McMurdo Station where it pays tribute to his contributions to marine science while also serving as a reminder of the dangers of working in Antarctica.

Yet another example of why researchers need to be careful when dealing with sea ice occurred at a field camp near McMurdo Station on November 14, 1987. It must have seemed like just another routine dive below the sea ice at New Harbor, a beautiful bay at the mouth of the Taylor Valley on the western side of McMurdo Sound. Mark McMillan, a certified research diver and undergraduate student at the University of California–Santa Cruz, was a volunteer on a research program investigating single-celled marine organisms known as *foraminiferans* (forams) that build calcium carbonate shells and live buried in the glacial silt of the seafloor. Antarctic forams are famous for their relative gigantism; they measure up to one-half inch across if you include their outstretched feeding appendages, or *pseudopodia* (false feet). Mark suited up and slipped into the dive hole for the twenty-first time of the season, carrying a forty-pound tripod secured to an array of floodlights that he planned to transport to the seafloor. Whenever scuba divers transport equipment underwater they use a lift bag, essentially a bag of air, to compensate for the object's weight at depth. The idea is to ensure that the object—in this case the array of floodlights—remains neutrally buoyant. Mark's lift bag was at first too small for the job, so he got back out of the ice hole and added more lift to the floodlights.

Steve Alexander, Mark's dive partner, looked up from the seafloor to see Mark descending toward him. Shortly after returning his attention to the photographs he was taking, Steve noticed Mark's floodlights lying near him on the seafloor. He turned his head upward and was shocked to see Mark ballooned up under the sea ice, his dry suit full of air and his regulator hanging free of his mouth. Steve ascended rapidly and with superhuman effort managed to push Mark over to the dive hole, releasing enough air from Mark's suit to free him from the sea ice and adjusting the level in the suit so he did not sink to the bottom. Once Steve and the dive tender got Mark out of the dive hole, they administered mouth-to-mouth resuscitation and oxygen until a helicopter arrived. Sadly, even with the helicopter corpsman administering CPR and oxygen, doctors back at the station were unable to revive Mark.

Later, dive officers investigating Mark's death found no problems with the valves on his dry suit. They surmised that Mark must have added air to his suit during the descent to intentionally maintain his neutral buoyancy. When he accidentally dropped the heavy underwater floodlight array, the additional air he had added to his suit would have turned him into a cork, jettisoning him toward the surface in what is known among scuba divers as an extremely dangerous "uncontrolled ascent." Divers with Mark's advanced level of training know how to dump air out of their dry suits to slow themselves by pulling the suit's seal away from the neck to rapidly release a large volume of air. The dive officers speculated that Mark's collar may have been forced upward by rapid changes in his suit pressure, squeezing his neck and triggering a carotid sinus squeeze, a condition caused by receptors in the carotid artery that signal the heart to slow down, leading to dizziness and blackout. Unconscious, he would have drowned. Mark, like Jeffrey, had a bright future ahead of him in Antarctic marine biology. The UCSC Friends of Long Marine Laboratory established the Mark McMillan Memorial Fund to support college students in their studies of marine biology. Some of these students have gone on to pursue graduate studies in Antarctica.

I was involved in a diving incident that was, in retrospect and comparison, much less serious. I was tending a dive for two young fellows on the research team who were collecting marine invertebrates below one of our dive huts off Cape Armitage in front of McMurdo Station. As is the general practice when dive tending, I noted the exact time that my two colleagues slipped through the ice hole and sank out of sight. In the subfreezing waters of McMurdo Sound, a typical dive lasts about thirty minutes. Indeed, as decompression diving is against U.S. Antarctic diving regulations, the length of an ice dive is determined more by chilled extremities than by limited air supply. If a diver were to experience a dry-suit failure (a serious flood of cold water into the suit), he or she would have to decide between doing a decompression stop on the way to the surface (to prevent the bends) and likely dying of hypothermia, or going straight to the surface without a decompression stop and getting the bends. One remarkable exception to the chilled extremity doctrine was Norbert Wu, a world renowned underwater photographer who worked with our research team one year. Norb could stay comfortably in the icy water for over an hour. My heart used to go out to Norb's dive buddy who would pose in the pictures to give scale to his stunning *National Geographic* photographs. The poor fellow would emerge shivering and blue. I awaited the divers' return, upon which my duty was to assist them with unclipping their various hoses and straps, and then hoist their weight belts and tanks up into the hut. Once freed of this considerable weight, with a combined scissor-kick and push up, each diver would shoot up, spin around, and come to rest on their rear end with their finned feet dangling in the water.

Thirty minutes had passed since my colleagues had descended, and yet there was no sign of their return. The first clue that a return is imminent is the appearance of bubbles in the dive hole. At first, these bubbles are tiny, like carbonation escaping a soft drink, making a delicate fizzing noise as they surface. Then, as the divers approach, the bubbles swell to the size of walnuts, each making its own popping sound. Once the divers have fully ascended, the ice hole is transformed into a noisy, frothing cauldron of fist-sized bubbles. At first, my colleagues' absence beyond a half hour was not particularly concerning. The divers did have plenty of air, and they may have just underestimated the time they needed to make collections. After forty minutes into the dive, I began to consider my options. I decided that if the divers had not surfaced after fifty minutes I would radio the station communications center, Mac Ops, and report a dive emergency.

My mind continued to work feverishly, running through every imaginable scenario. What could be holding them up? Then, with only a minute or two to spare before my planned emergency call, I had an idea. I lay down on my stomach, braced my arms against the floorboards, and poked my head as far down into the dive hole as I could without falling into the water. Peering below the water's surface, I saw the answer off to the side of the hut: a large Weddell seal was swimming in tight circles around the circumference of the dive hole. I jumped up and dashed out the door to our tracked vehicle, where I grabbed a long wood-handled dip net out of the back. Returning to the hut, I lay down next to the hole and, extending the wooden handle, gently but firmly poked the seal in the rear end. Startled, the seal quickly darted off.

Immediately, bubbles began to appear in the dive hole—then more bubbles, tons of bubbles. The clank of a scuba tank against the sea ice and the loud hiss of an air hose heralded the return of the first of the two divers, spitting saltwater and wiping his face. "Thanks!" he yelled. "We were just about to draw straws to see who was going up first." Weddell seals, unlike their distant cousin the leopard seal, are not dangerously aggressive predators; however, perhaps because they live in such close harmony with the sea ice, they love our dive holes, and sometimes choose to guard them. Divers who have surfaced through an ice hole coveted by a circling Weddell seal have occasionally been nipped in the crotch. So for the final twenty minutes of their dive, my colleagues had been anxiously waiting for the seal to depart, neither wanting to risk gaining an octave in their voice.

I have encountered Weddell seals in other odd scenarios as well. I've opened many a dive hut door to be greeted with the gut-wrenching pungent odor of putrid, semi-digested fish, indicating a seal's proximity. On several such occasions, I've even discovered the perpetrator of this horrendous stench fast asleep on the dive hut floor. Weddell seals can sleep deeply. I recall one that didn't wake up even after I noisily opened the door and stepped into the hut. I took a seat on the hut bench and quietly listened to the seal's deep, rhythmic breathing. Eventually, I, too, had lain down. If not for the pungent odor, I suppose I also might have fallen asleep.

A colleague of mine, Ken Dunton, was studying marine phytoplankton below the Arctic sea ice with a group of three other marine biologists. Each field season, the biologists commuted by helicopter daily to an insulated hut erected over a dive hole in the sea ice. One morning, the biologists noticed that nose-sized holes had been scratched in the inch-thick ice that had formed overnight. They discovered new holes each morning for months without the slightest view of the perpetrator. Then one day on a solo dive, Ken dropped into the dive hole and found himself face-to-face with a ringed seal, the most common sea ice–associated seal in the Arctic. These small Arctic seals rarely grow longer than four feet. Their name originates from their distinctive dark gray spots, each surrounded by a white ring. Ken described the encounter as the equivalent of two ten-year-olds exploding onto a playground on the first day of spring. The ringed seal zipped around Ken, performing somersaults and shooting through his legs. The two played games. Ken discovered that if he blew air bubbles, the seal would chase them as they ascended and push them along the underside of the ice with its nose. Ken's forty-five-minute dive flew by in delight.

*One of the most interesting features* of the seafloor in the southern coastal regions of McMurdo Sound is the conspicuous absence of seaweeds. Seaweeds, like most plants, need sufficient light to fuel the photosynthesis that produces their simple sugars. In many areas of coastal southern McMurdo Sound, two factors limit the availability of light. The first is that the annual sea ice breaks up later in the austral summer than it does further to the north. Some years, the annual sea ice may not break out at all, allowing it to thicken and for snow to build up on its surface. The second factor, well known to those who spend the winter at McMurdo Station, is that about three months of the year are spent in perpetual darkness—the sun never reaches above the horizon. While such a long stint in the dark is a challenge to humans, both physically and psychologically, it completely prevents the survival of any seaweeds. Conditions are different just fifteen miles to the north of the station at Cape Evans, where the light conditions under the sea ice are better. Here, the annual sea ice breakouts occur earlier in the season and more frequently, which allows just enough light to support some seaweed life. Two lone red seaweeds

(*Iridaea cordata* and *Phyllophora antarctica*) manage to hang on in this most southern fringe of their biogeographic range. To expand our team's studies of chemical defenses of Antarctic marine invertebrates to seaweeds, Chuck Amsler, Bill Baker, and I would drive our track vehicle up the coast to Cape Evans when the annual sea ice was suitably thick and dive to collect samples of the two red seaweeds.

Our team's collections of red seaweeds at Cape Evans revealed a remarkable story involving their chemical defenses. Chuck Amsler recounted his findings of an Antarctic feeding triangle featuring the red seaweeds growing at Cape Evans and two very common predatory marine invertebrates: sea urchins and sea anemones. Chuck first demonstrated that the red seaweeds defended themselves chemically against the sea urchins—they discarded bite-sized tissue pieces placed directly on their mouths but did not discard seaweed tissues whose distasteful chemicals had been removed by soaking the tissues in a chemical solvent. This was a clear demonstration that the seaweeds were chemically defended against the sea urchins. Chuck had noticed during his numerous dives that while sea urchins did not consume the red seaweeds they often picked up and carried pieces of drift seaweed on their spiny backs. Even in some areas where red seaweeds did not grow, sea urchins carried red seaweed that had drifted in from other locations. Chuck placed sea urchins in a shallow seawater table and presented them with equal amounts of shells, small rocks, and red seaweeds. The sea urchins always preferred to cover themselves with red seaweeds. Chuck then offered seaweed-covered and seaweed-free sea urchins to large predatory sea anemones. Those sea urchins with red seaweed on their backs escaped. Those sea urchins without red seaweeds were captured and eaten by the sea anemones. Red seaweeds could be a sea urchin's best friend. By then removing the chemicals from the seaweeds, Chuck found that the sea anemones could not care less about the seaweed's chemical defenses; they just couldn't get a grip on a sea urchin when it was covered with seaweed.

So who are the winners and losers in this Antarctic feeding triangle? The sea urchins clearly benefit because they are protected by the seaweeds from hungry sea anemones, and the seaweeds benefit because sea urchins carry reproductive plants into regions where seaweed populations are nonexistent. Here, their spores might populate new habitats. The sea anemones are the clear losers in this triangle because they lose out on a potential meal; however, overall, our casual scuba-surveys of the seafloor indicate sea anemones appear to be doing just fine. Maybe their success can be attributed to their eating just about anything that comes their way, including the occasional Antarctic jellyfish that swims by just a little too close to the seafloor.

Compared to McMurdo Sound, the central and northern coast of the western Antarctic Peninsula is the Garden of Eden. Here, a dense underwater forest of brown, red, and green seaweeds carpets the seafloor to a depth of about one hundred feet. Approximately 150 species of seaweeds take advantage of the increased sunlight provided by the reduction in the duration of annual sea ice. With light less constrained, the peninsular seaweeds flourish in the nutrient-rich Antarctic waters, some displaying remarkably rapid growth despite the frigid temperatures. If one were to randomly drop three-by-three-foot frames onto the seafloor, then collect and weigh the seaweed in each frame, the amount would be similar to that in the rich kelp forests along the coasts of the western United States and Canada.[2]

Brown seaweeds are the most dominant among the three color groups. The largest is *Himantothallus grandifolius*, comprised of a single massive blade that drapes thirty to forty feet across the seafloor. Tan starfish and brown limpets adorn its blades like ornaments dangling from a Christmas tree. Kelp like *Cystosphaera jacquinotti*, whose numerous small buoyant floats, called *pneumatocysts*, suspend the plants ten to fifteen feet off the seafloor, act as a sort of forest canopy. When coastal storms churn the sea, these buoyant plants sway back and forth in an elegant slow-motion dance. The understory is largely comprised of two closely related brown seaweeds, *Desmarestia anceps* and *Desmarestia menziesii*. Their branches rest prostrate on the seafloor or stand suspended by currents in the water column in lengths up to ten feet. Below the undulating brown seaweeds are delicate, finely branched red seaweeds such as *Plocamium cartilagineum*. Standing two to three feet high, their branches swarm with small amphipods. Green seaweeds are less conspicuous, but in deeper water the stringy green *Lambia antarctica* occurs, its branches constructed of a series of siphon-like tubes. Unable to compete for space among the plants, a nevertheless rich community of sponges, soft corals, hydroids, and sea squirts is relegated to vertical rock walls or to depths below which there is too little light to fuel the seaweed population. Yellow, orange, and red starfish perch, stomachs extruded, on sponge prey. Tiny snails, some no bigger than the head of a pin, and bottle-cap-sized limpets graze on films of microbes and filamentous microalgae. Like small manatees, black rock cod chew on seaweeds to compliment their diet of amphipods, snails, and limpets, their mottled-brown color rendering them almost invisible on the forested seafloor.

At first blush, scuba diving in open water and in undersea forests near Palmer Station sounds easy compared to diving below the sea ice in McMurdo Sound: no drilling or blasting is required and researchers can dive from zodiac boats. On occasion, researchers will do a shore-based dive when the weather is too rough for boat operations, either to do a check-out dive to ensure they are comfortable and proficient in the water upon arrival at the station or to collect local seaweeds or marine invertebrates. The downside to open-water diving is that local winds

near Palmer Station are blustery and downright unpredictable, sometimes even interminable, making small boat operations a challenge.

Diving among the picturesque seaweed forests comes with some
necessary precautions that divers must consider, as seaweeds can entan-
gle hoses or tank valves. Precautions typically include carrying a dive
knife to cut tough seaweeds that cannot be pulled free or broken by hand
and avoiding swimming through particularly thick forests of seaweeds.
My postdoctoral mentor John Pearse had a close call while diving in a
kelp forest off Carmel Point in California. He was so enthralled with the
crystal-clear water and giant kelps that when his dive partner signaled
that he was low on air and surfacing, John remained on the seafloor to
take advantage of the great conditions. Figuring that he could quickly
and easily make a free ascent from twenty feet, he sucked his air tank
almost dry. But when he headed toward the surface, the valves on his
tank became entangled in the canopy of kelp strands and blades. Panic
would have engulfed most divers, and John later admitted that in the
moment he wondered if he would break free. Instead, despite not having
his dive knife with him, he kept his cool and reached around behind his
air tank to pull a single strand of the entangled kelp toward his mouth,
removed his air regulator, and bit through the tough strand of plant tissue. Repeating this maneuver several more times, he was able to free
himself and surface safely.

Due to climate warming, the surface seawater temperatures along
the western Antarctic Peninsula have climbed a full 1 to 2 degrees
Fahrenheit over the past half century.[3] This temperature increase may
not sound like much, but when one considers that for thousands of years
the seawater temperatures along the Antarctic Peninsula have ranged
from about 30 to 34 degrees Fahrenheit, this two-degree increase represents a significant change for the marine life. In some respects, underwater forests may actually benefit as seawater temperatures rise because

warming seas mean less sea ice, and less ice means more light penetrating the water to facilitate photosynthesis. The underwater forests are likely to extend farther offshore and south along the western Antarctic Peninsula. At present, a little less than 3 degrees latitude, or about 150 miles south of Palmer Station, the seafloor at the British station Rothera has a much sparser seaweed community. In spite of the greater numbers of icebergs scouring the seafloor due to warming, these modest seaweed communities should gain a foothold as climate change further reduces the ice cover.

Rapid climate warming may also bring about more fundamental changes in Antarctic seaweed communities. Bill, Chuck, and our marine chemical ecology research team have demonstrated that most ecologically important Antarctic seaweeds have chemicals called secondary metabolites (because they play no role in metabolic processes) or natural products (because they are chemicals made by living organisms in nature) that make them distasteful to varying degrees to small grazing crustaceans and seaweed-eating fish and starfish. But what surprises even scientists is that the level of chemical defense exhibited by Antarctic seaweeds is as high as that found in tropical seaweeds where intense herbivory is notorious. So why are Antarctic seaweeds so well chemically defended? One of my doctoral students, Yusheng Huang, answered this question through a series of experiments. He would collect seaweeds from the seafloor near Palmer Station in fine-mesh bags and count the amphipods he found. By the time Yusheng had finished his doctoral studies he had accumulated 78,415 amphipods in his seaweed bags. Maggie Amsler, a marine crustacean biologist by training and an indispensable member of our expeditions to Palmer Station, then determined that these amphipods, or "pods" as we came to affectionately call them, included an impressive thirty-eight different taxonomic groups. Combining this information with Chuck Amsler's survey of the common seaweeds near Palmer Station, we estimated that in some cases as many as three hundred thousand amphipods surrounded individual seaweeds. The question of why the Antarctic seaweeds were chemically defended had been answered: as Yusheng's study had demonstrated, they were living in a virtual soup of hungry, fly-sized, crustaceans.

So what are the myriad of Antarctic amphipods consuming if not the large, chemically defended seaweeds? The answer literally lay within the direct object of this question. Large seaweeds harbor tiny seaweeds that live embedded within their tissues. Such *endophytes* (plants within plants) produce tender, threadlike filaments that emerge from the tissue of their host seaweed. Diving among the forests of Antarctica, a seasoned seaweed biologist would immediately notice the conspicuous absence of filamentous seaweeds, either growing on their own or emerging from host plants. Where were they?

Subsequent studies by Chuck and his doctoral student, Craig Aumack, solved the riddle. When Chuck and Craig grew Antarctic seaweeds in outdoor seawater tanks at Palmer Station some with and some without amphipods, the seaweeds lacking the presence of amphipods quickly sprouted threadlike filaments. In a month, filaments covered the seaweeds so thickly that one could barely see blades or branches. Further studies indicated that amphipods love to eat these tiny, tender filamentous seaweeds, gobbling them up as quickly as they emerged from the surfaces of their hosts. This story is marine ecology with a win-win twist. The amphipods gorge themselves on the filaments emerging from the larger seaweeds, and by doing so they help out those seaweeds by keeping their surfaces free of the smothering coat that would quickly hamper photosynthesis and growth. To add another ecological twist, Jill Zamzow, a National Science Foundation Postdoctoral Fellow who worked with our Antarctic team at Palmer, discovered that the large, chemically defended seaweeds provide protection for amphipods from predatory fish. The amphipods can hide in the interstices of the plants that fish won't bite into during a chase.

Rapid climate warming may also play into the chemical ecology of Antarctic seaweeds. Chemical defenses (secondary metabolites or natural products) come at a cost—their production requires energy. As warming diminishes the sea ice and seaweeds are exposed to greater levels of energy-producing light, the plants could invest these additional energy resources into the production of more chemical defenses. If this happens, it's anyone's guess as to who might be the winners and losers. One possibility is that the large ecologically dominant seaweeds would become chemically resistant to the filament-producing endophytes that live within their tissues. Such a scenario would eliminate the main food source for the voracious amphipods and, ironically, force the amphipods to consume the larger, chemically defended seaweeds—the proverbial monkey wrench thrown into the spokes of the marine community. Should large seaweeds come under attack by amphipods, the undersea forests would likely diminish and threaten critical habitat that provides refuge and food for a myriad of sea creatures ranging from the smallest of bacteria to the largest of fish.

Another potential challenge for seaweeds experiencing rapid warming is exemplified in seaweeds that occur in yet another global-change hot spot—the Pacific coast of southeastern Australia. My friend and colleague Peter Steinberg, who is krill biologist Debbie Steinberg's brother, is a marine microbial ecologist and the director of the Sydney Institute of Marine Sciences. Peter explained to me that rapid climate warming is causing a devastating bleaching disease that is showing up in large, ecologically dominant Australian seaweeds. The disease causes the seaweeds to lose the pigments that both color them and play a key role in photosynthesis. Infected portions of the plants bleach bright white, which compromises the health of the entire plant by reducing energy production. Peter and his colleagues have linked the bleaching disease to changes in naturally occurring films of bacteria—slime-producing microbial communities known as *bacterial biofilms*—that attach to and generally grow harmlessly upon the surfaces of healthy seaweeds. As climate change causes seawater temperatures to rise, warming stresses natural biofilm-forming bacteria and allows pathogenic bacteria to infiltrate the biofilm and break down the seaweed's pigments.

Bacterial biofilms such as those growing on the surfaces of Antarctic seaweeds are amazing living communities. Many biofilm bacterial species exhibit a type of decision-making process called *quorum sensing* that allows individual bacteria to coordinate their behaviors with other species. Quorum sensing gives bacteria forming healthy biofilms on the surfaces of seaweeds a leg up to protect themselves from pathogenic bacteria. Imagine thousands of fans filling the stadium at a football game preparing to do "the wave." As the individual fans assemble, they communicate with one another by raising both arms over their heads. Accordingly, each fan can assess when their overall numbers have reached the critical mass necessary to successfully carry out "the wave." Bacteria in biofilms coordinate with one another in a similar fashion, only rather than using arm signals to communicate, they use signaling molecules called *pheromones* or auto-inducers. Bacterial quorum sensing was first discovered in *Vibrio fischeri*, which lives in the photophores (light-producing organs) of Hawaiian squid. Both organisms mutually benefit from the presence of *Vibrio fischeri:* the bacteria acquire a protective retreat as well as nutrients from their squid host. Squid exploit the added bioluminescence of the bacteria in displays of aggression or seduction. When sparse amounts of *Vibrio fischeri* exist on their own, they do not luminesce, but when they populate a squid's photophore, they become highly concentrated, and their genes produce *luciferase*—a molecule that catalyzes a light-producing chemical reaction.

The marine invertebrates of the Antarctic Peninsular forests—sponges, sea squirts, soft corals, starfish, sea urchins, sea slugs, and many others—are also vulnerable to the rapidly rising seawater temperatures. These creatures are most sensitive during the early stages of life. Whether parents protect (brood) their embryos or cast the embryos into the sea to develop as tiny swimming larvae, a rapid increase in seawater temperature is likely to have serious consequences. Antarctic marine invertebrates and their offspring are generally unable to cope with even modest changes in temperature—a condition known as being *stenothermal*. (*Steno* is derived from Greek, meaning "narrow," so *stenothermal* refers to a narrow tolerance for temperature change). This sensitivity is not surprising considering that frigid Antarctic seas have remained at an almost constant temperature for millions of years. With even a modest 3 or 4 degrees Fahrenheit increase in the seawater temperature, Antarctic embryos and larvae may either die outright or develop too fast. Normally, Antarctic embryos and larvae develop at a record-breakingly slow pace, some two to three times slower than the development of similar species living in temperate or tropical seas. So why would it be risky to develop faster in a warming world? The answer is simple: it would disrupt feeding patterns. If the rates of development of Antarctic larvae that feed on seasonal blooms of phytoplankton are accelerated, they will, in a sense, show up at the cafeteria weeks before it opens and would thereby starve. Antarctic larvae that do not feed on the plankton and are well-provisioned with an infusion of maternal nutrients would be less susceptible to this scenario, but they too might find themselves crawling about on the seafloor too early to take advantage of the nutritious by-products that drift to the seafloor during plankton blooms.

Adult marine invertebrates are similarly not immune to climate warming, and the few experiments that have been performed indicate that temperature increases may exact a cost. In a simple but elegant experiment, Lloyd Peck,[4] a distinguished professor from the British Antarctic Program, exposed Antarctic limpets and scallops to current and anticipated seawater temperatures and measured their behavior. He found that when he flipped limpets over in seawater at the current temperature of about 32 degrees Fahrenheit, they righted themselves within twenty-four hours. However, when the temperature was increased to 36 degrees Fahrenheit, only 50 percent of the limpets eventually turned over. The more drastic outcome turned out to be true for Antarctic scallops: those held at the normal temperature swam by clapping their shell valves when touched with a probe, while those exposed to 36 degrees Fahrenheit couldn't swim at all. Further studies have found that even when temperatures are raised more gradually, many Antarctic marine invertebrates simply cannot adapt to the increase. Lloyd Peck and David Barnes, another highly respected British Antarctic marine biologist, surmised that future warming of sea temperatures to levels only as high as 4 degrees Fahrenheit above current maximum temperature would pose long-term survival issues for populations of marine invertebrates on the shallow Antarctic shelf.[5]

Some may argue that the climate in Antarctica has changed in the past and that marine invertebrates will simply adapt to a warming sea. The problem with this argument is that evolutionary adaptation generally takes place over hundreds of generations and over much longer periods of time. Anthropogenic climate change is warming Antarctic seas at unprecedented rates. Antarctic marine invertebrates can have extraordinarily long lifespans when compared to most temperate and tropical counterparts. Sea urchins and scallops can live over a century and sponges many hundreds of years. One specimen of the Antarctic sponge *Cinachyra antarctica* was estimated to have lived 1,500 years.[6] Over the course of their individual lives, these marine organisms might experience a temperature rise that challenges their ability to cope.

The option of *acclimation*—adjusting one's physiology to changes in environmental conditions, a capacity found in marine invertebrates in warmer seas where temperatures are more variable—is less of an option for Antarctic species that rarely deal with a changing environment. Nonetheless, for Antarctic marine invertebrates with biogeographic distributions that extend northward to the subantarctic and that maintain regular gene flow through larval dispersal, sufficient genetic variability may ensure thermal resilience and offer some hope of recolonization should less thermally resilient populations be eliminated by the rapid warming. Despite an urgent need to expand upon our knowledge of how Antarctic marine organisms—plant and animal—respond to rising temperatures across all their life stages (young and old) and species ranges (narrow and wide), in the big picture the biodiversity of the Antarctic shelf is vulnerable to predicted warming. The consequences of a loss of biodiversity could encompass everything from altering key Antarctic marine food chains to the loss of species that may hold cures to cancer.

*Icebergs are unique in shape, color, and texture. This one is well over one hundred feet tall with eight hundred feet of ice that remains hidden below the sea. With glaciers and ice sheets rapidly breaking up, more icebergs will flow into the Southern Ocean. Photo credit – Russell Manning*

*The Marr Glacier, adjacent to Palmer Station on Anvers Island, calves a large chunk of ice into Arthur Harbor. When I first began coming to Palmer Station twelve years ago, these glacial calving events were infrequent. Now, the harbor next to the station regularly resounds with the thunderous roar of ice crashing into the sea. Photo credit – Bill Baker*

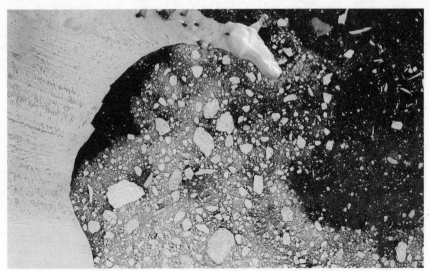

*This NASA satellite image captures the dramatic breakup of the Larson-B Ice Shelf in 2002, which previously measured the size of Rhode Island. At least eight other ice shelves have disintegrated over the past thirty years. With less resistance, glaciers adjacent to former ice shelves are flowing more rapidly into the sea and accelerating the rise of global sea levels. Photo credit – NASA*

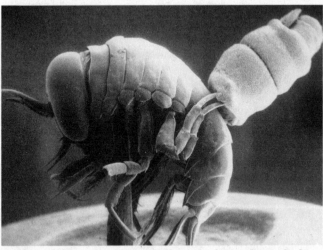

*Antarctic waters are teeming with zooplankton, about which scientists know very little. Fish biologist John Janssen and I discovered this tiny amphipod measuring a quarter of an inch. These amphipods capture and carry sea butterflies because the butterflies harbor a distasteful chemical that deters fish predators. Photo credit – Philip Oshel*

*Wearing "dry suits," my dive buddies and I (far left) prepare to descend through a hole drilled through six feet of sea ice in McMurdo Sound, Antarctica, to collect marine organisms from the seafloor. Back in the 1980s we used regulators with two hoses to feed air to the mouthpiece. Today, dive suits are state of the art and single-hose regulators are the norm. Divers carry a backup single-hose regulator in case of an emergency. Photo credit – James McClintock*

*The seafloor is covered with marine algae, sponges, soft corals, sea squirts and sea slugs, and other shell-less invertebrates that contain chemicals that have been shown to be active against deadly flu viruses and melanoma skin cancer. Photo credit – Bill Baker*

*These clam-like organisms called brachiopods share the seafloor with sea urchins and a sea cucumber. Clams, scallops, brachiopods, snails, and other shelled Antarctic marine invertebrates are highly vulnerable to ocean acidification—known as "the other $CO_2$ problem"—which is caused by the absorption of atmospheric carbon dioxide by the world's oceans. Photo credit – Bill Baker*

*Graduate students Julie Schram (left) and Kate Schoenrock (right) are testing the impact of ocean acidification and rising seawater temperature on marine algae and invertebrates. In just a few decades, the carbonate building blocks used by Antarctic marine invertebrates to make their shells will become scarce. Photo credit – James McClintock*

*The U.S. National Science Foundation research vessel* Nathaniel Palmer *cruises through sea ice off the coast of Antarctica. During this December 2009–January 2010 expedition, Sven Thatje directed our field team. Using the SeaBed, an unmanned automated submersible vehicle, the team discovered large populations of invading king crabs. Photo credit – Sven Thatje*

*Three adult king crabs in Marguerite Bay on the lower western Antarctic Peninsula. The largest crab on the rock is about twenty inches from leg tip to leg tip. As sea temperatures rise, king crabs are invading shallow shelf waters and pose a threat to sea stars, brittle stars, sea urchins, clams, and snails. Photo credits – Richard Aronson and James McClintock (Project Principal Investigators)*

*I'm in front of a former Adélie penguin rookery (areas with smaller stones) on Torgersen Island in February 2010. Forty years ago there were about 15,000 breeding pairs on this and several surrounding islands. Today we're down to 2,500. Snowmelt from unseasonal snowstorms is drowning the eggs, and a severe reduction in annual coastal sea ice makes the long swim to edible krill too hard. Photo credit – Jason Cuce*

*A dense colony of chinstrap penguins nesting on eggs at Aitcho Island in the South Shetland Islands about seventy-five miles north of the Antarctic Peninsula in December 2009. As the climate rapidly warms, chinstrap and gentoo penguins are establishing new breeding colonies along the western Antarctic Peninsula. Photo credit – James McClintock*

*A leopard seal yawns while resting on an ice floe along the western Antarctic Peninsula. Leopard seals—eight- to ten-foot-long top predators that prey on penguins and seals— are typically found where there is sea ice. Their survival is at risk with climate warming along the Antarctic Peninsula. Photo credit – Christopher Srigley*

*The hand-drawn "Get Well" poster for my son Luke, who was in an automobile accident, from everyone at Palmer Station. The image was delivered to Luke at the hospital the next day as a two-by four-foot poster. I am holding one end of the sign; Maggie Amsler is holding the other end, and Chuck Amsler can be seen standing on the bottom on the staircase with his arm on the railing. Photo credit – George Ryan*

*Chuck and Maggie Amsler pose in front of Amsler Island in 2008. Four years earlier, the "island" lay buried under the tongue of the glacier and was considered a point of land. New islands are emerging from under the glacial ice as glaciers rapidly recede up and down the length of the Antarctic Peninsula. Photo credit – Charles Amsler*

*The stunning mountains of the Antarctic Peninsula in the vicinity of Wilhelmina Bay on the Danco Coast of the northwestern Antarctic Peninsula. The mountains that stretch the length of the Antarctica Peninsula are an extension of the Andes in South America. The image captures the unparalleled grandeur of the landscape. Photo credit – Richard Harker*

# Chapter 5

# Polar Acid Seas

*T*he most common sea butterfly in the seas of Antarctica—the lovely *Limacina antarctica*—is poised for catastrophe. The same Southern Ocean that has for eons provided both habitat and nourishment is turning acidic and threatening its survival. Its shell, thin as a human hair and measuring half an inch across, is sculpted of elegant whorls of translucent mineralized aragonite. Ocean acidification—which can damage or dissolve these thin, fragile shells—is the enemy. What these swimming snails give up in body size they make up for in sheer numbers. A single bathtub of Antarctic seawater contains up to five thousand individuals. A cubic mile of the surface of the Southern Ocean houses 27 trillion butterflies. Their mind-boggling abundance renders this mid-sized planktonic organism (mesoplankter) a classic keystone species. The production and decay of their shells is of sufficient magnitude to play into the global carbon cycle—a combination of biological, chemical, and geological processes involving the exchange of carbon between all of the world's natural systems, including soils, rocks, lakes, ocean, and atmosphere. In the Southern Ocean, the construction of countless shells of sea butterflies requires the uptake of carbon-rich aragonite, one of two common crystal forms of calcium carbonate. Upon the snail's death, the shells slowly sink, transferring the carbon within them to the deep seafloor. Once they reach the bottom, the shells either dissolve or form fine silt sediments called *pteropod ooze* that ultimately become rock and clay.

Shelled sea butterflies such as *Limacina antarctica* have unique reproduction and feeding habits. When they are young, the butterflies are all males. As they age and approach sexual maturity, they change into females, an intriguing transition known as *protandric hermaphrodism*. But if this were not odd enough, just before changing, males mate with males, each storing away the mutually exchanged sperm until it has completed the final transition to becoming a female. As females, the snails then use the stored sperm to fertilize their eggs and release the resultant embryos into the sea where they develop through two phases of swimming larvae. First-stage larvae are shaped like spinning tops bristling with tiny hairs they use for swimming. Second-stage larvae are tiny versions of juveniles-to-be, complete with shells but equipped with a thin flap of tissue covered with tiny hairs called a velum that they use for swimming and capturing food particles. When a larva prepares to metamorphose into a juvenile, it absorbs its velum, providing nutrients and energy for this final stage of development. As juveniles, shelled sea butterflies launch into a lifetime of herky-jerky swimming generated by rapid flaps of wings. After several months, fully grown and with mature reproductive organs, they join the rank of adults. Surviving a year or more, Antarctic sea butterflies are long-lived compared to their temperate and tropical cousins.

Their feeding strategy is as odd as their sex life. Secreting a mucous web, they suspend themselves in the water column like parachutists, the vast expanse of the webs dwarfing their individual body size. Shaped like a sphere, the webs serve to provide buoyancy and capture prey. Like fishers, the shelled butterflies periodically reel in their sticky webs, consuming any prey entangled in the mucous. Shelled sea butterflies are quick to abandon their elaborate webs when disturbed. As such, vast numbers of deserted webs drift in the plankton and serve an important ecological role as breeding grounds for bacteria and as food for smaller planktonic organisms. With shelled sea butterflies as numerous as stars in a galaxy, even their feces become important in the web of life. The sinking fecal pellets play a key role in transporting organic material from the surface of the Southern Ocean to its deepest depths.

Victoria Fabry, a biological oceanographer, is an expert on ocean acidification and its impacts on shelled sea butterflies. In 2007, Vicky and several of her students traveled to McMurdo Station to study what happens to the shells of Antarctic sea butterflies when they are exposed to conditions of elevated ocean acidification. The team placed the shells of sea butterflies into seawater tanks adjusted to the acidity and mineral levels predicted to occur in Antarctic seawater by the year 2100. In less than a week, the central coils of the delicate translucent shells turned opaque and they began to dissolve. At three weeks, the shells were completely opaque, and at one month, the once-smooth symmetrical coils of the shell had become crooked and ragged. After six weeks, the investigators ended their experiment—the once-intact shells were but ghostly shadows of their former selves, no longer suitable domiciles for sea butterflies. Over a span equivalent to a tenth of their life expectancy, the shells of the sea butterflies were no more.

In a second dramatic illustration of the vulnerability of polar sea butterflies to ocean acidification, scientists used a powerful atomic microscope to observe cracks and valleys in the micro-architecture of the outer surfaces of shells of a subarctic sea butterfly. Remarkably, these living individuals had been placed in water that mimicked conditions of future acidified seawater for just forty-eight hours. Yet sea butterflies face not hours but a lifetime in an acidified sea. Polar shelled sea butterflies are in peril. In addition to their shells transporting carbon to the seafloor, they play a critical role in food webs—their flesh providing essential nourishment for animals further up the food chain, including a vast array of jellyfish, fish, marine birds, and baleen whales. Should sea butterflies become unable to produce or maintain their fragile shells in a future ocean, the prospects for the myriad sea creatures that exploit this bountiful buffet are bleak.

Scientists have coined ocean acidification "the other $CO_2$ problem," a problem I discuss in Chapter 1, which has yet to receive the same level of attention that newspapers, magazines, television, and the Internet have paid to the inescapable role that carbon dioxide plays as a potent greenhouse gas—the original $CO_2$ problem. But its back-burner status is changing. Marine scientists are voicing their collective concerns about the potential impacts of ocean acidification on natural ecosystems and fisheries, and popular science writers are taking note. Far-reaching educational media outlets such as *National Geographic* are featuring timely articles such as "The Acid Sea" by Elizabeth Kolbert.[1] The answer to why our oceans are growing increasingly acidic has its roots in the Second Industrial Revolution (late nineteenth century to present) that brought fundamental transitions in agriculture, manufacturing, mining, technology, and transportation. Birmingham, Alabama, where I live, is a poignant example of this revolution. In the late nineteenth century through the early to mid-twentieth, Birmingham, rich in natural resources like iron ore, coke, and limestone, exploded as one of the greatest steel producing regions in the world. Smoke and particulates from the foundries became so thick that drivers had to turn on their headlights in downtown Birmingham during daylight hours. The inevitable spread of industrialization around the globe led to expanded exploitation of oil, coal, and natural gas, that, in concert with deforestation (fewer trees to absorb carbon dioxide), elevated levels of carbon dioxide in our atmosphere to unprecedented heights. Because the oceans exchange gases with the atmosphere, what humankind puts into the air eventually ends up in the ocean. Scientists now estimate that the world's oceans have absorbed about 30 percent of the more than five hundred billion tons of carbon dioxide released into the air since the onset of the Industrial Revolution.[2]

The chemistry behind ocean acidification can be simply expl
When seawater absorbs atmospheric carbon dioxide, it immediate
acts with water, releasing hydrogen ions that lower the pH of the seawater. The pH scale provides a measure of the concentration of hydrogen
ions and thus the level of acidity. The lower the pH value the higher the
acidity. Since the beginning of the Second Industrial Revolution, the
average pH of the world's oceans has been reduced from pH 8.2 to pH
8.1.[3] This decrease does not sound like much, but a decline of one-tenth
of a pH unit in human blood can cause acidemia, a condition that can
lead to coma and death. This similar small decline in the pH of seawater
actually represents a 30 percent increase in the acidity of the world's
oceans. Scientists predict that if humans continue to burn fossil fuels
at the current rate, by the end of the twenty-first century the world's
oceans will have an average pH of 7.8, making them a stunning 150 percent more acidic than at the onset of the Second Industrial Revolution.[4]
Unfortunately, the amount of time required to undo the damage of ocean
acidification is considerably greater than the time necessary to cause
it. Even if by the end of the century humankind were to cease expuls-
ing carbon dioxide into the atmosphere, it would take tens of thousands
of years for the pH of the world's oceans to return to pre–Industrial
Revolution levels.

A second important outcome of atmospheric carbon dioxide react-
ing with water is that a significant portion of the hydrogen ions released
combine with dissolved carbonate ions in seawater to form bicarbon-
ate ions, thus depleting the global pool of oceanic carbonate ions—the
building materials used by a myriad of sea creatures to manufacture
their calcium carbonate skeletons or shells. Thus, the world's oceans,
saturated with the carbonates aragonite and calcite, are now predicted to
become limited in carbonates over the next fifty to one hundred years.[5]
The dwindling carbonates will challenge the ability of marine organisms

to build and maintain their skeletons and shells. Among calcified marine organisms on the Antarctic Peninsula shelf, those competing for the declining pool of carbonates include coralline red seaweeds, both flattened and erect forms hardened by a skeletal matrix; soft corals, which lack the external skeletal cups of reef-building hard corals but have skeletal needles (spicules) that provide rigidity; clams and snails, which generally have thin outer shells; brachiopods, shelled and clam-like, but with unique filter feeding structures called lophophores; and starfish, brittle stars, crinoids, sea cucumbers, and sea urchins, with skeletons of microscopic or platelike ossicles and spines.

Household examples of the impacts of exposing skeleton and shell to acidic solutions help conceptualize ocean acidification. Drop a human tooth into a cup of carbonated soda and try to find it a few days later—you won't. Submerge an uncooked chicken egg in a bowl of vinegar for twelve hours—the hard shell will disappear, leaving only a paper-thin vitelline membrane. Add the shell of a common garden snail to the vinegar, and twenty-four hours later you can see the erosion on the surface of the shell with a magnifying glass. And it turns out that more than skeleton and shell are vulnerable to the impacts of an acidic ocean.

Scientific studies with marine invertebrates have found that embryonic development, growth rates, muscle mass, immunity, sensory capacity, and even aspects of behavior are all at risk. Even vertebrates such as fish have their problems. A seminal study by Philip Munday, a tropical fish biologist at James Cook University in Australia, demonstrated that colorful orange-and-white striped clown fish lose their ability to find their way home when exposed to conditions of ocean acidification.[6] These fish cannot discriminate between chemical odors when placed in seawater at a predicted end-of-century pH of 7.8. Rather than navigating toward familiar scents indicative of their home territory, the clown fish moved toward smells from objects that were unfamiliar—seaweeds and other marine organisms outside their normal environment—making them easy prey. When Munday and his colleagues decreased the pH level to 7.6 in the experimental tanks, the clown fish became completely lost, no longer able to follow the most simple odor trails. *Olfaction* (smell) is important to clown fish as soon as they hatch from their eggs and swim away from the reef and out to sea. After twelve days, they follow waterborne odors to return to a reef to find an anemone willing to provide a safe retreat. Ocean acidification may completely change how clown fish behave directly from birth.

Despite an abundance of treelike soft corals, shallow Antarctic seas lack several key ingredients to support their reef-building hard cousins, including warm temperatures and sufficient year-round light to support photosynthesis. At first, it may seem odd that an animal like a coral requires light, but the coral's skin houses thousands of tiny plantlike organisms that use the sun to make energy. These tiny plants provide the majority of the coral's nutrients and energy. The corals, in turn, provide the plants' refuge and essential elements such as nitrogen and phosphorus. This relationship is a wonderful example of mutualism—two organisms living together, each benefitting from the other. Nonetheless, despite Antarctica lacking conditions suitable for hard corals to harbor energy-producing plants in its shallows, they do hide deep below.

Deep-water hard corals occur on the continental slope and the deep seafloors surrounding Antarctica. Widely distributed in oceans, these corals occur to an abyssal depth of eighteen thousand feet. Until the end of the twentieth century, studies of deep-water hard corals were only possible when fishing boats accidentally caught them in their nets. Even today, while working with deep corals is possible by submersible, it is technically difficult and expensive. As such, little is known of their biology except that they use their tentacles to capture plankton for food rather than depend on symbiotic photosynthetic plants living in their tissues. Hard corals form deep-sea reefs amid the featureless soft sediments, providing a critical habitat for rich assemblages of microbes, sponges, hydroids, bryozoans, anemones, starfish, sea urchins, and fish. These reefs are the rain forests of the deep. Deep corals are also useful barometers of climate change—when they produce their calcium-carbonate skeletons they record a chemical signature of the seawater. Paleoecologists drill through deep coral reefs and use the cored chemical information to reconstruct patterns of deep ocean circulation. This information is important to gauge how human-induced climate change may impact vital deep ocean currents—currents that, in turn, influence global climatic events such as continental-scale floods, droughts, heat waves, and deep freezes. The Southern Ocean that surrounds Antarctica is arguably one of the most important oceans in terms of influencing climate patterns because it provides the main connection between the planet's ocean basins and the lower and upper layers of global ocean circulation.

Exactly how deep-sea hard corals in Antarctica and elsewhere will respond to ocean acidification is unknown. Studies to evaluate potential impacts are difficult to carry out given the immense labor of collecting or manipulating corals at such great depths. One possibility is to study the few deep-water hard corals found at scuba-diving depths in the fjords of Alaska and Chile. These unique populations may provide marine scientists a window into the biology of deep-water corals. In the meantime, marine scientists can examine the potential impacts of ocean acidification on shallow-water hard corals. Their findings so far paint a collage of outcomes—some species, stressed by a combination of rising temperature and ocean acidification, will perish outright; others may be able to survive ocean acidification by increasing their rates of calcification to offset a dissolving skeleton, and still others may survive despite losing their skeletons entirely.

In a remarkable study, two Israeli marine scientists, Maoz Fine and Dan Tchernov, exposed living colonies of the hard corals *Oculina patagonia* and *Madracis pharensis* to seawater adjusted to levels of anticipated future ocean acidification. They provided the coral colonies enough light to satisfy their photosynthetic symbionts. After one month, the scientists noticed that the coral polyps exposed to ocean acidification had grown taller and that skeletons encasing each individual polyp had dissolved. The two colonial corals were, in a sense, no longer colonial. Rather, they consisted of a collection of "naked" individuals attached to a rock. The shell-less polyps survived in the experimental tanks for an entire year. The scientists transferred the coral polyps back to seawater at normal oceanic pH where, in an interesting twist, the naked polyps recloaked themselves in their calcium carbonate skeletons. While this scenario offers some hope of survival for some hard corals in the face of acidification, Fine and Tchernov were quick to point out that the decalcification of future coral reefs would be disastrous. The ecological implications of deconstructing a reef would be profound—reduced biodiversity, depleted food resources, lost habitat, etc.—and a tremendous amount of benefits coral reefs provide to human society would decline; examples include the fishery industry, recreational activities, and tourism.[7]

A major ecological player second only to corals in shallow hard-coral reefs is the encrusting red coralline algae (seaweed). Hard as rock, coralline algae deposit mineralized calcite within their cell walls. These hard, flat plants are the mortar of the reef, cementing corals and other organisms into a sturdy matrix resistant to daily tidal currents and the ferocity of episodic storms. Studies indicate that encrusting coralline algae are even more vulnerable to the impacts of ocean acidification than hard corals. In an elegant examination of the impacts of ocean acidification on calcified algae, marine scientists from federal agencies and universities in Florida, Bermuda, and Hawaii set up six large outdoor experimental tanks (called *mesocosms*) near a Hawaiian reef and then pumped unfiltered seawater from the reef into each tank. Within each tank, the scientists placed clear plastic cylindrical containers to provide surfaces on which settling spores of marine algae could attach. Bubbling carbon dioxide into the seawater in a subset of the mesocosms, they acidified it to a level predicted to occur by the end of the century. Over the next seven weeks, the scientists carefully followed the settlement and growth of young seaweeds on the cylinders in each tank. In the end, the differences between the seaweeds in the tanks with normal pH seawater and the tanks with acidified seawater were striking. Encrusting coralline algae were practically nonexistent in the acidified seawater tanks, while heavy settlement occurred in the natural seawater tanks. Encrusting corallines that were able to settle in the acidified seawater grew slowly compared to those in normal seawater. The acidified seawater tanks also contained more noncalcified algae than the normal tanks. In summary, this scientific study revealed that future reefs may shift to soft, noncalcified algae, and the few hard corallines that survive will grow slowly, reducing reef growth. With fewer coralline algae to provide an architectural mortar, future coral reefs will become fragile and susceptible to storm damage.[8]

Ocean acidification could also pose a challenge to the livelihood of red coralline algae in the undersea forests along the western Antarctic Peninsula. Much of what scientists know about its distribution and abundance at Palmer Station can be attributed to Chuck Amsler, who is an expert on algae. Chuck was deployed by the National Science Foundation to Palmer Station in 1989 as part of a "rapid-response team" of marine scientists whose mission was to evaluate the environmental impacts of the largest fuel spill to date in Antarctica. The spill had occurred when an Argentine resupply ship, the *Bahia Paraiso*, plowed into a reef in front of Palmer Station on January 28, 1989.

Chuck arrived a month later. He and his colleagues conducted a series of scuba dives to videotape underwater transects that had been placed by the divers at different depths along the seafloor. The examination of a range of depths was important because each type of seaweed is fine-tuned to exploit a specific spectrum of light to produce energy (sugars) from the process of photosynthesis. Accordingly, by sampling at different depths, Chuck and his colleagues could ensure that they sampled a wide diversity of seaweeds. Despite having initially killed and oiled some sea birds, most of the high-grade fuel evaporated quickly. Chuck and his colleagues' dives quickly revealed that the spill had not impacted the algal communities.

But Chuck realized the work he and his colleagues had conducted could be used for further scientific study. The video he had in hand was also the first detailed assessment of seaweeds living near Palmer Station. They revealed a rich algal community comprised of both fleshy, noncalcified algae and hard coralline seaweeds. The tapes also showed that encrusting corallines, which are susceptible to ocean acidification, covered about three-quarters of the rocky surfaces on the seafloor. Clearly, encrusting coralline seaweeds were an important ecological player in Antarctic Peninsula marine communities. However, because coral reefs are absent in shallow Antarctic seas, these encrusting coralline algae were not serving as mortar to cement hard corals together. More likely, the textured bumps and ridges of the corallines provide habitat for a diversity of tiny plants, animals, and biofilms that attract and sustain settling larvae. Marine biologists have long known that red coralline algae give off unique water-borne odors that serve as "settlement inducers" for marine invertebrate larvae, including those of abalone (large flat-shelled snails) and sea urchins. Laboratory studies have shown that larvae encountering odors of red coralline algae rapidly settle to the surfaces of aquaria and metamorphose into juveniles. In nature, the scents of coralline algae may function to enhance settlement of offspring in areas where rocky substrates provide refuge and food.

In 2009, our University of Alabama at Birmingham research team conducted the first investigation of the potential impacts of future ocean acidification on Antarctic encrusting coralline algae. At the time, we did not have the luxury of working with living coralline algae in Antarctica (a grant from the National Science Foundation allows us to do so throughout the field seasons in 2012 and 2013). Instead, we took advantage of an earlier opportunity to make collections while several members of our team had been at Palmer Station working on aspects of marine chemical ecology. Divers collected limpet shells with and without encrusting coralline algae, knowing that the latter often encrust the upper surfaces of the former. Maggie Amsler shipped the shells back to our laboratory in Alabama. We adjusted the seawater to a pH of 7.4 (an aggressive level of ocean acidity predicted to occur by 2300) by bubbling carbon dioxide into beakers filled with seawater. A similar set of beakers contained normal pH seawater. We dropped Antarctic limpet shells with and without encrusting coralline algae into the beakers. By the end of two weeks, the shells of the limpets in the acidified seawater weighed less than those in normal seawater—they were already dissolving. By the end of eight weeks, the skeletons of the encrusting corallines on limpet shells submerged in acidified seawater had lost twice the weight of coralline skeletons in normal seawater. Moreover, the corallines were cracking apart, suggesting they might peel right off the limpet shells. If live Antarctic coralline algae peeled off hard surfaces, the algae would be dealt a fatal blow; it is unlikely they would survive to reproduce. When we returned to Palmer Station in 2012, we set up laboratory experiments to answer whether living coralline algae exposed to ocean acidification will compensate for their skeletal loss by producing additional calcium carbonate (results were still forthcoming at the time of publication).

Barnacles, on the other hand, have a shell design that compensates particularly well for an increasingly acidic ocean. These calcified filter-feeding crustaceans glue themselves to hard surfaces to secure a resting spot for life. Thomas Huxley, a nineteenth-century British naturalist, anatomist, and philosopher, and others have famously described the barnacle as a "little shrimplike animal, standing on its head in a limestone house, and kicking food into its mouth."[9]

These odd crustaceans are ecologically important because they compete for living space with other encrusting species and provide food for higher organisms in coastal rocky habitats that are exposed at low tide and submerged at high tide. Michelle McDonald, a graduate student in my lab, was interested in examining the impacts of ocean acidification through the life cycle of barnacles. Raising these creatures is both a science and an art: one has to know how and where to collect sexually mature adults, gather and feed the larvae, and convince them to settle and grow. Rather than attempt to collect and raise them ourselves, we turned to the world's authority on barnacles, Professor Dan Rittschof, a zoologist at the Duke University Marine Laboratory in North Carolina, to culture larval barnacles and expose them to natural seawater or conditions of ocean acidification. Once the larvae metamorphosed into juveniles, he would ship them live back to our university in Alabama so that we could examine potential impacts of ocean acidification on their growth through adulthood and egg production.

The Duke Marine Laboratory sits on picturesque Pivers Island, surrounded by estuaries, tidal sand bars, and inlets, and within the barrier islands of the Outer Banks. Just across the Beaufort Inlet and within sight of the marine lab is the historic coastal town of Beaufort. Michelle, Chuck Amsler, and I met Dan's research associate, Beatriz Orihuela, who walked us down to the lab dock to collect *Amphibalanus amphitrite*, a common, dime-sized intertidal barnacle. Lying flat on the dock, Beatriz used a hammer to pound the encrusting adults off the wooden pilings, then dropped them into a bucket of fresh seawater. Nothing was delicate about this operation—especially when the barnacles that had been hammered off the pilings were subject to another stiff round of pounding on the bottom of the bucket. But what Beatriz was doing with the hammer actually made sense—the crushed shells liberated thousands of tiny naupliar larvae, baby crustaceans with three pairs of arms and a single eye. Placing the bucket in a darkened room, Beatriz trained a powerful beam of fiber-optic light at the water's surface on one side of the bucket. Attracted to the light, the swimming nauplii gathered in swarms. Beatriz repeatedly sucked up the thick soup of nauplii with a turkey baster and squirted them into beakers filled with filtered seawater. Over the next month, Michelle nursed ten glass beakers containing barnacle larvae held in either natural seawater or seawater acidified with bubbling carbon dioxide gas. She changed the water and fed the larvae a daily ration of phytoplankton. In each beaker, Michelle recorded the timing of transition to the cyprid larval stage, the sizes of the cyprid larvae, and the timing of juvenile metamorphosis and settlement on the sides of the glass beakers. Her studies revealed that larval barnacles do just fine under the conditions of future ocean acidification. But all was not as rosy in Michelle's subsequent studies of adult barnacles.

The live juvenile barnacles attached to their beakers arrived in a large Styrofoam cooler. Moistened only by paper towels soaked in seawater, they were quite happy—an advantage of working with a marine invertebrate accustomed to exposure to air during low tides. Michelle filled the beakers with either normal pH seawater or seawater with the pH lowered with bubbling carbon dioxide. Over the next several months, Michelle measured the barnacles' growth, and when they became reproductive adults, she counted the bright yellow eggs she saw through their transparent basal discs. She found that ocean acidification influenced neither growth nor reproduction.

At the end of the experiment, however, an examination of the adult shells revealed that the basal discs—the component of the shell glued to the substrate and thus not exposed to seawater—of barnacles raised in acidified seawater were larger and had higher levels of calcium carbonate than those from barnacles raised in normal pH seawater. In other words, the barnacles responded to ocean acidification by producing additional basal disc shell material. But the wall shells—comprised of six plates that make up the barnacle's outer shell—from barnacles that Michelle had raised in acidified seawater told a different story. Although they were similar in size to those from barnacles in normal pH seawater, they were much easier to crush than wall shells from barnacles raised in normal seawater. Despite the barnacles bolstering the calcium carbonate in their shells in the presence of acidified seawater, they were unable to stop the portions of their shells that were exposed to acidified seawater from dissolving. In ecological terms, this shell degradation means that predators living in future oceans may find it easier to crush or drill into barnacles. Should this increased susceptibility translate into fewer barnacles, then less food for higher predators would be available and encrusting communities in rocky intertidal habitats around the world could look quite different than they do today. Despite their small size, barnacles offer a model to study the general impacts of ocean acidification on all marine organisms that produce shells. Because parts of their shells do not come in contact with seawater, scientists using molecular tools can use this to their advantage to understand how certain genes are "turned on" to generate additional shell material when some organisms are exposed to increasingly acidic seas.

As with climate warming, Antarctica is in many respects the earth's most well-suited natural laboratory to study the "first effects" of global ocean acidification due to two key factors. First, Antarctica has naturally low concentrations of carbonate ions—attributable to both carbon dioxide dissolving easier in ice-cold seawater and to unique patterns of the Southern Ocean's currents mixing with one another. As a result, lower levels of carbonates such as aragonite and calcite, with which creatures build a skeleton or shell, occur in Antarctic seawater. Second, the skeletons and shells of Antarctic marine organisms are generally very weakly calcified. I have picked up Antarctic clams and snails and without so much as a gentle squeeze accidentally crushed them in my hand. Some shells are so thin that one can see right through them. This may be, in part, the result of calcification requiring more of an energy investment at low temperature or of the evolutionary outcome of living in an environment with no crushing predators. Most likely, the result is a combination of both factors. The bottom line is that weak-shelled Antarctic marine invertebrates living in seas that already lack the skeletal building blocks they need will be the first and foremost to face the consequences of an increasingly acidic sea.

The field of studying Antarctic ocean acidification is so young that marine scientists are scrambling for even the most basic information about how vulnerable calcified marine organisms actually are. My beloved echinoderms—starfish, sea urchins, brittle stars, feather stars, and sea cucumbers—comprise such a group of marine invertebrates. Echinoderm skeletons are made of magnesium-calcite, a mineral even more vulnerable to ocean acidification than pure calcite, or, especially, aragonite. The higher the ratio of magnesium to calcite in an echinoderm skeleton, the more vulnerable it is to dissolving. Skeletons of temperate and tropical echinoderms examined to date have levels of magnesium sufficient to classify them as highly vulnerable. Is the same true of Antarctic echinoderms? A review of the scientific literature indicates

that the skeletal mineral composition of only a single Antarctic echinoderm is known to science.[10]

How can scientists evaluate the vulnerability of Antarctic echinoderms to ocean acidification if they don't even know their magnesium-
calcite levels? Jason Cuce, a graduate student working with Bill Baker
during the 2009 field season at Palmer Station, helped to make up for
some of the dearth of information. Due to some equipment issues, Jason's
chemical ecology research project had drawn to a premature completion,
which meant Jason was available to join a team of biologists collecting
fish for their physiology studies aboard the *Gould*. Led by Professor Bill
Detrich of Northeastern University in Boston, the biologists caught a variety of fish by either deploying baited traps or towing a large net called a
*trawl* along the seafloor behind the ship. The trawl often surfaced with a
healthy variety of starfish, brittle stars, and sea cucumbers, and on occasion, sea urchins, and feather stars in its mesh. In the spirit of generosity
so legendary among Antarctic scientists, Professor Detrich extended a
warm welcome to Jason to join on a series of weeklong fishing cruises to
collect these echinoderms so the team could find out just how vulnerable
their skeletons were to ocean acidification.

Jason returned from these excursions with a wide variety of samples for skeletal analysis, and these were augmented with collections
made by divers near Palmer Station. They were shipped frozen to us
at the University of Alabama at Birmingham, where they were thawed
for skeletal analyses and dissected. A strong solution of bleach disintegrated the various body parts, rendering sparkling clean skeletons that
were subsequently shipped to a laboratory in Canada for mineral analysis. The results came back several weeks later: every single one of the
twenty-six species in our study fell into the "high magnesium-calcite"
category. Antarctic echinoderms were indeed highly vulnerable to ocean

acidification. Among the groups of echinoderms examined, the brittle stars and starfish had the highest ratios of magnesium to calcite and were, therefore, the most vulnerable to an acidic sea. This heightened vulnerability is important because brittle stars, and especially starfish, are keystone predators whose habits influence how Antarctic seafloor communities are structured. Removing them from the equation would destabilize the food chain.

Ecology aside, more practical implications of ocean acidification strike at the heart of the global seafood industry. Shelled mollusks such as clams and oysters are particularly vulnerable to decalcification. Mark Weigardt and Sue Cudd, local owners and operators of a shellfish hatchery in Tillamook, Oregon, had dedicated sixty years between them to ensuring that oysters were available at local restaurants, seafood markets, and oyster bars. In 2007, the oyster larvae at Mark and Sue's Whiskey Creek Shellfish Hatchery began to die. Mark and Sue first considered pathogenic bacteria a likely culprit, but when scientists found no bacteria, they engaged a team of oceanographers to identify the problem. They discovered that the seawater being pumped to their hatchery from nearby Netarts Bay was so acidic that it was preventing shell growth in the larvae. Now that they knew what the problem was, the oceanographers began to notice a trend. When north winds created upwelling conditions, whereby deeper, more acidic seawater flowed into the pipes that fed the hatchery, the oyster larvae died.

Similar die-offs of oyster larvae had occurred in other bays along the West Coast. Shellfish growers had used bays as natural nurseries to grow oysters since the 1920s, and the fishery (an entity engaged in raising or harvesting shellfish) now produces millions of dollars of oysters per year. At first, the larval die-offs had been written off as just a few bad years, but when oyster larvae died in 2006, 2007, and 2008—at hatcheries up and down the Pacific coast, from Puget Sound to Los Angeles—the trend became alarming. Now, the oyster fishers are asking Congress to help replumb seawater sources to the hatcheries and to provide monitoring equipment so that the fishers know when they need to take precautions against acidified seawater.

The pressure on Antarctica's fisheries is already growing. Given that Antarctic seas are vulnerable to the first impacts of ocean acidification, the stresses on harvested species may intensify. The Antarctic cod (also called the Antarctic toothfish) occurs in the ice-laden waters surrounding the Antarctic continent and is sold in fish markets and restaurants along with its cousin, the Patagonia toothfish (marketed as Chilean sea bass) that lives along the Pacific coast of Chile and south to Cape Horn, the Falkland Islands, South Georgia, and west of the Antarctic Peninsula. The Antarctic cod and Patagonia toothfish are both fished legally and il-legally. At present, nothing is known about the potential impact of ocean acidification on Antarctic cod. But just as clown fish become disori-ented and eventually dysfunctional when exposed to ocean acidification, one cannot rule out behavioral or physiological consequences of ocean acidification on Antarctic cod. The societal implications if ocean acidifi-cation negatively impacts cod populations worldwide are profound. Cod is stuffed into just about every fish sandwich in every cafeteria and fast food joint on almost every continent. Overfishing caused the legend-ary crash of the Canadian and American Atlantic cod fisheries in the early twentieth century; humankind can ill afford to yet again squander a valuable food resource. Even the smallest risk of impacting present day cod fisheries is reason to pause and consider the impacts of the ever increasing acidity of the world's seas. Should commercial fishing con-tinue for Antarctic cod, ocean acidification may be the least of its prob-lems. During my seven field seasons at McMurdo Station in the 1980s and 1990s, I came to know Professor Art DeVries, the discoverer of the Antarctic fish antifreeze (a glycoprotein compound). This remarkable compound binds to miniscule ice crystals within the blood of Antarctic fish, preventing the blood from forming a solid latticework of ice. The natural antifreeze makes it possible for Antarctic fish to undergo the remarkable evolutionary process of *adaptive radiation*—whereby through a recent single ancestor a rapid divergence in body design generates new species that occupy a variety of ecological niches (to date scientists have identified 122 species of Antarctic fish). One can picture an ecological niche as an organism's address and what it does for a living. Just as Darwin observed a variety of finches that had evolved different body designs (beak shapes) so as to exploit different habitats and food resources, adaptive radiation, whether acting on fish or finches, produces an array of species occupying a variety of habitats. Art has spent more than forty field seasons at McMurdo Station and has caught and released more than six thousand Antarctic cod for his study, evaluating their population biology. Art's population studies are geared to answer a variety of interrelated questions, which include: How long do Antarctic cod live? How old are they when they reproduce? How rapidly do they grow at different stages of their lives? How far do they venture? These are all critical questions to answer before fishery biologists can make sound decisions about fishery management practices. In recent years, a modest fishery for Antarctic cod in the southern Ross Sea has resulted in less samples for Art to study. With slow growth and sexual maturity not occurring until ten years of age, Antarctic cod are highly vulnerable to overfishing and should become a protected species.

Currently, no fisheries in Antarctica exist for bivalve shellfish such as clams, oysters, or scallops. Scallops would be a tempting target for an unscrupulous fishing industry because large populations of *Adamussium colbecki* occur along coastal regions of the Antarctic continent. I once witnessed vast scallop populations while scuba diving in New Harbor on the eastern shore of McMurdo Sound, with densities of up to thirty adult scallops in a single square yard. Besides the potential impacts of ocean acidification, Antarctic scallops grow so slowly that a fishery would not be sustainable. Still considered young at ten years of age, adult scallops may live as long as one hundred years. The scallops would be fished out before they ever had the chance to reproduce. Crustaceans are also considered "shellfish," and *Euphausia superba* (two-inch-long shrimp-like crustaceans) have been commercially harvested by various nations since 1980. These Antarctic krill are freeze-dried, ground into a power, and used primarily to extract omega–3 fatty acid dietary supplements or in livestock feed and pet food. These krill are high in fluoride and not particularly palatable, and thus humans consume only a small portion of the catch. The USSR dominated the krill fishery during the 1980s and early 1990s, netting and processing on average about three hundred thousand tons each year before dropping out of the business. Since then, Japan has led the way with annual catches of about sixty thousand tons. Six other countries, including the United States, have played lesser roles in the Antarctic krill fishery.

A recent study published in 2010 by krill biologist So Kawaguchi and his students at the Australian Antarctic Division in Tasmania reported that krill embryos and larvae exposed to a moderate (pH 7.7) level of ocean acidification showed no effect on krill embryonic development or larval behavior. But at a higher level (pH 7.4) only 10 percent of the krill embryos hatched, and not a single larva survived. While the higher level of ocean acidification used in their experiment was aggressive, the investigators point out that the eggs of *Euphausia antarctica* are spawned at the sea's surface and then sink 2,100 to 3,000 feet before hatching and swimming back to surface waters. As such, krill eggs and larvae are exposed to deeper, more acidic seawater.[11] Importantly, the large body of scientists who coauthored the 2007 International Intergovernmental Climate Change Report predicted that continued oceanic absorption of carbon dioxide will lead to a global seawater pH of 7.6 within the depth range of Antarctic krill by the end of the twenty-first century.[12] Studies are urgently needed to refine the seawater pH level at which the impacts of ocean acidification will first be felt by Antarctic krill. And the Antarctic krill fishery is currently under pressure from other fronts. Concern is growing among some Antarctic marine scientists that the parties to the Convention on the Conservation of Antarctic Marine Living Resources (CCAMLR) that regulate fisheries in Antarctica have set too high a fishing quota for Antarctic krill. Recent studies indicate that populations of *Euphausia antarctica* along the northwestern half of the Antarctic Peninsula and the Scotia Sea (the region between the Antarctic Peninsula, South Georgia, and Tierra del Fuego) have declined dramatically over recent decades because of rapid climate-induced shifts in the marine ecosystem.

Agencies around the globe have recognized how important it is to understand the impact of ocean acidification. I was invited to participate in the first meeting between all the key U.S. organizations—the National Science Foundation, the National Oceanic and Atmospheric Association, the U.S. Geological Survey, and so on. American scientists of all kinds—from biological oceanographers concerned about the fate of the world's plankton, marine ecologists examining the fate of coral reefs, marine chemists and geochemists deciphering the impacts of acidification on the chemical and mineral properties of seawater, shellfish biologists lamenting the decimation of oyster hatcheries, and even environmental sociologists deliberating about how best to educate the public—delivered fact after fact pertaining to the growing risks that ocean acidification poses to our oceans.

# Chapter 6

# The March of the King Crabs

*T*he spiny king crab paused, resting its massive walking legs. Its reddish exoskeleton, bristling with sharp spines, is a remarkable architectural feat of evolution. Despite the need to be periodically shed, this thin outer shell provides a rigid framework for the attachment of muscles and ligaments engineered to connect, pull, and retract the skeletal elements that coordinate the movements of the legs and a large pair of highly maneuverable claws. The crushing capacity of these two appendages is formidable, strong enough to crack a walnut shell. Similar to the claws of its cousin the stone crab, which are a popular seafood treat, each claw exerts a tremendous force powered by the synergy between striated and smooth muscle, the former rapidly contracting, the latter acting much more slowly but stronger and for a sustained period. In tandem, these muscles could pinch each claw devilishly tight, the tooth-lined ridges used to draw blood, tear tissue, crack bone, and crunch thick shells. Watching the screen of the monitor, Sven Thatje, a lead scientist aboard the British research vessel *James Clark Ross*, saw a long pair of flexible antennules project below the king crab's stalked eyes. Moving back and forth through the seawater, the crabs sensed the familiar plume of chemical odors originating from their compatriots. Thirteen crabs in all, some male, some female, and each measuring about two feet across, had been migrating as a group over the fine silt sediments in search of prey on the deep seafloor of the Southern Ocean. Several weeks earlier they had encountered what was at first a gently rising slope. Here, they began an epic climb toward Antarctica, which hadn't happened for millions of years. As seawater temperatures warm with rapid climate change, king crabs are no longer confined by low temperatures to the deep and may reach a point where they would emerge from the deep onto the Antarctic shelf. Because of patterns of currents, sea temperatures on the Antarctic slope and shelf have traditionally been lower than temperatures of the neighboring deep sea. As king crabs are vulnerable to low temperature (they are unable to regulate important ions in their blood), populations have been unable to move up into the colder slope and shelf waters. Now, as sea temperatures warm, that is changing.

As the days passed and the slope steepened for the thirteen, the king crabs may have sensed that they were moving beyond their normal deep-water foraging range where meals are sporadic compared to shallower depths with more dependable food resources. Nonetheless, the crabs had little inkling that they were entering a realm from which they had been physiologically excluded for millions of years. Their inability to properly balance magnesium ions in their blood at low temperature had precluded them from advancing upon the Antarctic slope and shelf. Now, rapid, human-driven climate change was warming the Southern Ocean.

Like wolves, spiny king crabs prowl the world's deep seafloors in search of a smorgasbord of starfish, brittle stars, clams, snails, and other smaller crustaceans. Once the crabs corral and subdue a prey, they use their claws to hold it tightly and crack it open. A pair of mandibles rapidly tears and shovels pieces of tissue into their mouths. Ever opportunists, deep-sea king crabs also take advantage of *food falls*: carcasses of shrimp, fish, and even seals and whales falling to the seafloor. Gathering around corpses, the crabs join a host of worms, starfish, crustaceans, and other scavengers partaking in the feast.

Like most deep-sea invertebrates devoid of the seasonal cycles of light or temperature that serve as an internal clock to time reproduction, the king crabs reproduce year round. Sex between consenting king crabs includes a precoital courtship where males gently stroke and tap the females with their walking legs. When they finally mate, the males grasp the females and, using a pair of rear legs, paste their sperm in the form of a tidy little package, or *spermatophore*, onto the genital opening. Later, after the eggs are fertilized, the females gently gather them and hold them under an abdominal flap below their bellies. The tens of thousands of brooded offspring develop slowly in the deep cold conditions, sometimes requiring as much as two full years to hatch. When the crabs release the fat little larvae, they are packed full of nutrients. As such, unlike the larvae of most crabs, they don't need to feed as they swim just above the seafloor. In the deep sea, having their own internal source of nutrition and hiding out near the seafloor is advantageous to king crab larvae—they don't have to feed on the highly seasonal plankton and thus face less risk from planktonic predators. Eventually, the larvae shed their last exoskeletons in a long series of molts and settle to the deep seafloor.

*The spiny king crabs,* now rested, returned to their laborious climb. Their ascent had taken them well over eight thousand feet above the seafloor, some two-thirds of the distance to the shelf break, a transitional seafloor region about 1,500 feet below the sea surface where the slope abruptly flattens to become the Antarctic Shelf. Suddenly, the king crabs came to an abrupt halt. Unaccustomed to light, they were startled by the powerful beams of floodlights mounted on the top of the front frame of ISIS, the United Kingdom's premier deep-diving Remotely Operated Vehicle (ROV). The submersible vehicle, rectangular in shape and about eight feet long and four feet high, was, like the crabs, equipped with two manipulator arms. The ROV slowly approached the crabs, its propellers making only the softest of whining noises. The sub stopped about ten feet from the crabs. In complete silence, the thirteen king crabs stared at the submersible floating above them. The submersible, with its video camera fully operational, stared back. The date was January 25, 2007, and Sven was currently positioned 3,684 feet above in the Bellingshausen Sea off the coast of western Antarctica. He pointed excitedly at the image of the large king crab smack in the middle of one of the nine video monitor screens in the shipboard "control van" that houses the operational guts and the eight support personnel and pilots for ISIS. "Just as I suspected!" he exclaimed. The king crab images on the monitor screen suggested that the narrowest of temperature margins may have restricted crabs from climbing the Antarctic slope, for just a fraction of a degree of warming had opened the proverbial door—an invasion, Sven reasoned, whose ultimate outcome might very well be the colonization of the Antarctic shelf by large populations of king crabs.

*Although we'd exchanged emails* over a period of five or so years, I first met Sven face-to-face in 2009 in Melbourne, Florida, at the home of my friend and research colleague Rich Aronson. Sven is a principal scientist and lecturer in marine evolutionary biology at the National Oceanography Centre in the coastal port city of Southampton, England. He had been postulating for years that the large populations of king crabs residing in the deep waters of the Southern Ocean might soon begin to migrate up the Antarctic slope toward the shelf. As is the case of most well-grounded hypotheses, the basis of Sven's predictions owed its inception to the blending of two well-known facts. First, because of the inability of deep-sea king crabs to properly regulate magnesium ions in their blood at very low temperatures, they act as if they have just polished off several martinis in rapid succession when they're exposed to the cold. They stumble about, unable to walk, and most certainly are in no shape to feed or reproduce. These behaviors would impede the establishment of crab populations, and are almost certainly why crabs and lobsters, the two major groups of marine crustaceans with large crushing claws, have been unable to inhabit the cold polar seas that encompass the slope and shelf of Antarctica. Rich Aronson and other invertebrate paleontologists have noted the absence of crabs in their documentation of fossil records of marine invertebrate communities of Antarctica. Second, despite Antarctic seas having persisted at low temperatures since Antarctica separated from South America some forty million years ago, recent and rapid global climate warming has warmed the deep waters of the Antarctic slope ever so slightly.[1]

Sven, Rich, and I, all of us intimately familiar with marine communities of the Antarctic seafloor, quickly recognized that the impact of a large, deep-water crushing predator moving into the fragile marine invertebrate communities of the Antarctic shelf could be devastating. And if that were not enough, the extension of the population from the slope to the shelf appeared to be imminent; these thirteen crabs foretold an event that would likely be measured in years rather than decades or centuries.

For thirty minutes, the king crab had frozen, motionless, below the hovering ROV. Now, along with its compatriots, it returned to its methodical march, either stimulated to action in an effort to move away from the bright floodlights, or trusting that since there had yet to be an attack, the submersible was not a predator. Far above, the sub pilot nudged Sven, "They're climbing again, shall we bring one up?" "Absolutely," Sven replied. The two robotic arms on the front of the ROV, each with double-pronged insect-like pinchers aptly called "Isis's hands" by the sub's operators, whirred into action. Six independently controlled propellers, two for vertical movement, two for fore and aft movement, and two for horizontal movement, gave the submersible great maneuverability. Used in combination, these thrusters could rotate the sub into just about any position. Using a joystick, the pilot skillfully maneuvered the manipulator arms, requiring a considerable level of dexterity and situational awareness. The slightest miscue could result in a mechanical arm crashing into delicate science equipment, cameras, or floodlights. Slowly guiding the tethered submersible into the proper position, the pilot fully opened the pinchers and then extended one of the arms. The king crab was no match for the large pinchers, and it was placed into the milk crate–sized "biobox" on the lower front end of the ROV that stored live samples. As soon as the sub pilot secured the lid of the box, the submersible, along with its historic sample, began its ascent back to the ship waiting at the surface.

Detailed images of king crabs marching across the seafloor thousands of feet below the sea are the product of a long and challenging history of submersible technology. Mechanical, electrical, and computer engineers had to design underwater vehicles capable of withstanding pressures in deep water. In addition, the vehicles needed to be able to maneuver with great precision. Marine biologists working in tight collaboration with submersible engineers helped design cameras, floodlights, and mechanical devices with suction tubes and arms for collecting organisms. Over my career, I have had several opportunities to gather scientific data using automated unmanned vehicles (AUVs), ROVs, and even a manned submersible that took me on an unforgettable journey to a depth of 3,000 feet off San Salvador Island in the Bahamas.

*The thirteen king crabs that Sven discovered* climbing up the Antarctic slope played a pivotal role in catalyzing what has become a productive research collaboration between me, Sven, and Rich Aronson. Now that the scientists had discovered a few king crabs on the Antarctic slope, systematic studies were desperately needed to evaluate the extent of their presence in Antarctic waters. Good fortune came knocking in the winter of 2009 when the National Science Foundation (NSF) invited me to participate in a weeklong workshop for American and Swedish marine scientists at an old castle-turned-conference center hidden away in a pine forest just north of Stockholm, Sweden. The workshop would lay the groundwork for a collaborative U.S.-Swedish Antarctic research program that would begin upon the Swedish icebreaker *Oden* as it made its commissioned pilgrimage to the Ross Sea. For a relatively small investment, the two nations could engage in cutting-edge collaborative polar science by capitalizing on *Oden*'s five-year contract with the NSF to break out a channel in the sea ice for the annual austral summer resupply of McMurdo Station.

My sleeping accommodation at the workshop was not in the castle but was instead a tiny room in a cabin in the snow-covered woods, and our conference meeting room was in a neighboring convention building. Nonetheless, we ate our meals in the castle's restaurant with a view of an ice-covered lake through the large plateglass windows. Those of us at the workshop discussed at great length target areas of Antarctic marine science that would be suitable to study aboard the *Oden*, as well as the geographic regions that we would traverse during the cruise: the lower western Antarctic Peninsula, the Bellingshausen Sea, the Amundsen Sea, and the western Ross Sea. In the end, the scientists at the workshop agreed that despite the *Oden* being essential for clearing a path through the sea ice, the ship had limited infrastructural capacity to carry out science. This problem would require some creative solutions.

The thirteen-thousand-ton *Oden*, the size of an American football field, is the second most powerful icebreaker in the world. Its four diesel engines generate 24,500 horsepower (the equivalent of 106 BMW 325 sport sedans), giving it the ability to break through six feet of ice at a speed of three knots. Swedish engineers designed the *Oden* to break ice, not to conduct science. As such, the ship has no built-in research laboratories, no science support personnel, little in the way of scientific instrumentation, and perhaps most importantly, no propulsion system with the ability to "hold station." A pilot can position an oceanographic ship over a sampling site by deftly engaging bow and stern thrusters. Remaining on a fixed position can be critical while deploying sampling nets and various pieces of scientific equipment. Before embarking on the voyage, some of the limitations could be addressed; crane operators could hoist railroad car–sized containers, equipped as modular portable science laboratories, and secure them to the deck; oceanographic science technicians could be added to the crew manifest; engineers could bolt a hydraulically operated, triangular-shaped A-frame to the aft deck for deployment of cabled gear; and scientists could carry out some limited scientific sampling while the icebreaker was adrift or secured to the sea ice. But in the end, all of us at the workshop realized that much of the scientific potential of the research expedition would be compromised without additional infrastructure.

We found our solutions to the additional demands for research infrastructure in the *Nathaniel Palmer*. The NSF named the ship for the American sealer, explorer, and sailing captain Nathaniel B. Palmer, who may very well have discovered Antarctica. Many historians consider "Captain Nat" and his crew to be the most likely candidates for this honor, having landed on the Antarctic Peninsula during their 1820–1821 expedition to survey the Southern Ocean for sealing grounds. Remarkably, Captain Nat accomplished this historic feat in a paltry forty-seven foot sloop, the *Hero*, at the tender age of twenty-two. I am not sure who at our planning workshop had the means of garnering federal approval for deploying the *Palmer* in tandem with the icebreaker *Oden*, but they deserve a standing ovation. The two ships could now augment one another's capabilities. The *Oden* provided the brute force necessary to break through thick ice, while the *Palmer*, with half the horsepower of the *Oden* and only a modest capacity to break ice, had outstanding scientific research facilities built into its 308-foot length. Now, twice as many marine scientists could be accommodated on the two ships, and more importantly, the *Palmer* could deploy scientific equipment and collect samples while holding a fixed position. To further improve the expedition's scientific capability, a helicopter would be based on one of the ships, which would ferry marine scientists to and from the sea ice to study the ice's properties or to access penguins, seals, and whales, while the two ships could continue shipboard science activities. Departing Stockholm, our team had defined the parameters for an exciting and productive international collaboration. The next step would be for both governments to issue a formal call for research proposals, facilitating a competitive review process.

Roberta Marinelli, my program officer at NSF Polar Programs, called me as I stood on a rain-soaked sidewalk in downtown Portland, Oregon. She had promising news. The NSF review panel handling the joint U.S.-Swedish research expedition was recommending funding the grant proposal Rich Aronson and I had submitted to carry out the first systematic, broadscale search for king crabs along the coast of Antarctica. Although the review panel had placed our proposal into their "high risk" category, meaning they considered our chances of success modest at best, they realized that if we were successful, the implications would be profound. If recent climate warming were facilitating the movement of king crabs up the Antarctic slope, we could determine the extent of their movements, and we could also begin to evaluate the ecological implications. We would try to answer several important questions, including: Were the king crabs associated with particular types of communities? Were they mating? Were females carrying eggs? Were they consuming and thus impacting prey populations? The answers to these fundamental questions would allow Antarctic marine ecologists to determine if the king crab populations were becoming firmly entrenched and the extent to which king crabs would alter Antarctic marine food webs, and to make predictions about whether some species of prey consumed by king crabs might even go extinct.

Maggie Amsler, along with two graduate students—Roberta Challener, a doctoral student from my laboratory at the University of Alabama at Birmingham, and Steph Vos, a student working toward her master's degree with Rich Aronson at the Florida Institute of Technology—met up with Sven Thatje, who served as the chief scientist on the expedition, aboard the *Palmer* in Punta Arenas, Chile, in late November 2010. In addition, three ocean engineers under the leadership of Hanu Singh of the Woods Hole Oceanographic Institute National Deep Submergence Facility had joined the crab team to deploy and operate the submersible sled that we'd use to search for king crabs. To round out the team, two Danish graduate students, Rasmus Rasmussen and Sanne Kjellerup, from the laboratory of Per-Olav Moksnes, a marine larval biologist at the University of Gothenburg, Sweden, would sample for crustacean larvae in the plankton using a MOCNESS net (multiple opening/closing net and environmental sensing system) capable of collecting plankton at different depths.

These scientists had joined a host of others studying various aspects of the physical, chemical, and biological oceanography of Antarctica. One project led by Patricia Yager, an oceanographer from the University of Georgia, would examine the oceanography of Antarctic *polynyas* (Russian for "natural ice hole"), large areas of open water surrounded by broad expanses of sea ice that remain unfrozen during much of the year. Polar scientists suspect Antarctic polynyas have important ecosystem-level impacts, in that they provide a sunlit environment that enhances the growth of the tiny plants and animals comprising the phytoplankton and zooplankton, but little is known about them. Another project would examine the accumulation of toxins and pollutants in Antarctic seals by testing blood samples from Weddell and crabeater seals. All in all, about fifty scientists were involved in the international expedition.

Maggie and Sven knew their way around the port town of Punta Arenas, having passed through a number of times on their way to and from Antarctica. In contrast, Roberta and Steph were on their first voyage to Antarctica and had never been to Punta Arenas. So, before sailing, the two young women addressed several important traditions. First and foremost, they walked to the town square to pay homage to the statue of Fuegan natives seated at the base of Ferdinand Magellan. Maggie had directed Roberta and Steph to *rub* one of the native's big toes to ensure a safe return from their voyage. On my first visit, I mistakenly thought I was supposed to *kiss* the toe. I should have been clued in when I leaned toward the toe and noticed its polished sheen. Unfortunately, the sheen had not computed until after I had planted the kiss. I suspect a couple of tourists' photographs of that crazy American (me) kissing the native's toe are still circulating on Flickr. The evening before the *Palmer* departed for the Drake Passage, the ship's chief scientist had organized the traditional departure dinner for all the scientists on the cruise. The choice of restaurants was Sotito's, a favored dining spot for scientists coming and going from the Antarctic Peninsula. Everyone gathered around a series of small tables that restaurant staff had pulled together to yield one extremely long table, emblematic of collaboration and solidarity. Up and down the table, old and new friends alike raised a glass in a ceremonial toast with the traditional Chilean *Pisco Sour*—a limeade-tasting but insidiously potent mixture of grape brandy and lime juice. Pisco Sour stories are legendary.

The game plan was for the crab team to work aboard the *Palmer* for a month and then, with the exception of Rasmus and Sanne, to transfer to the Swedish icebreaker *Oden* for the balance of the two-month cruise. The team was able to address its primary objective at the onset of the expedition when the *Palmer* spent an entire week searching for crabs in Marguerite Bay, an extensive bay at the base of the western Antarctic Peninsula bordered to the north by Adelaide Island and to the south by Alexander Island. Sven had studied some preliminary evidence that invasive crabs might be in this region. The crab team's three-day voyage across the Drake Passage was smooth and uneventful. Unlike the *Gould*, the *Palmer* takes the seas well.

One year, I, too, had the good fortune of crossing the Drake Passage on the *Palmer*. I was the sole scientist on the trip that was headed to Palmer Station. My arrival turned out to be most inauspicious. The *Palmer*, too large to dock, had anchored in Arthur Harbor within three hundred feet of the station. In the excitement of the moment, I hurriedly stowed my gear aboard the zodiac. The morning was sunny and the sea calm but chock-full of small chunks of icebergs called bergy-bits and even smaller pieces of brash ice. So sure was I of my imminent arrival that I stood proudly in the bow of the zodiac as if I was George Washington crossing the Delaware River. At first, we made good headway toward the dock, yet as we moved closer to shore, the wind picked up and the ice shifted, impeding our progress. What appeared to be a five-minute jaunt now turned into a half an hour, then an hour. Having never imagined that my short zodiac taxi ride would require more than a few minutes, I neglected to put on sunscreen or to don my cold-weather clothing. My fingers and toes had grown cold, my ears numb. In the end, the short distance to shore took me a full two hours. This embarrassing episode was a poignant reminder of the guiding principle of polar survival. Never, ever, expect anything but the unexpected.

Upon their arrival at Marguerite Bay, the team immediately got to work. In consultation with the *Palmer*'s captain, Sven worked out a series of cruise-track transects that would span the length of the bay. While traversing each transect, the ship pulled the submersible sled at a steady speed across the Antarctic shelf and down the slope at a right angle to the shoreline. Despite towing the submersible at depths spanning 1,500 to 6,000 feet, the ship's winch operators could regulate the cable length so precisely that the sled remained at six to nine feet above the seafloor. The tethered submersible was a marvel of technology. Rectangular in shape and about eight feet long and four feet high, it was equipped with a digital camera system and strobe lights that snapped a photograph of the seafloor every three seconds. Each image was transmitted immediately to a computer and monitor on the mother ship. The crab team divided up their various tasks, some collecting and sorting plankton samples, others gathering information on seafloor temperature and depth. Sven, Maggie, Roberta, and Steph took turns watching the monitor in the ship's winch-control room. If they saw an image from the seafloor that they wanted to capture for immediate analysis, all they had to do was hit the save key on the computer keyboard. The three submersible engineers, John Bailey, Jeff Kaeli, and Frank Weyer, ensured that the computer operations and sled camera system worked smoothly. Over the next five days, the *Palmer* had traversed eight lengthy transects in the Bay, photographing an impressive one hundred miles of uncharted seafloor.

On December 2, the excitement was running high among the crab team as the submersible was lowered into the sea for the first time to begin its search for king crabs. The sled took an hour to descend, and the monitor screen revealed only the periodic jellyfish or squid emerging from the darkness of the water column. Then, suddenly, the craft was there, and the first image of the shelf seafloor was broadcast to the screen: a field of soft fine sediments with rocks scattered about. Within five minutes, a sponge appeared on the screen to the collective "ooohs" and "aaahs" of the team. The excitement mounted over the next few hours as blue fish, dark purple octopi, shrimp, red starfish, brittle stars, cream-colored anemones, and deep-sea corals came into view. Each animal triggered a lively discussion about its identity, behavior, and ecology. As the submersible sled dropped off the Antarctic shelf and began to descend the slope, the suspense was suddenly broken by a collective cheer as the first king crab appeared on the monitor. Sven was unable to control his joy. He leapt to his feet, spun around, and high-fived each of the members of the crab team. Later, Roberta recounted that seeing the first king crab had generated a sense of "pure elation" among the team. "It meant that this high-risk venture had paid off," she said.

The first clue that a "crossing ceremony" was afoot came in the form of an email. Each of the inductees was informed by a seemingly perturbed "King Neptune" that they had better get to work preparing a gala for the King in the ship's galley in several days' time. The Antarctic Circle, one of the five major circles of latitude on the earth, separates the true Antarctic Zone from the Southern Temperate Zone at latitude sixty-six degrees, thirty-three minutes and forty-four seconds south. The twenty or so scientists on board the *Nathaniel Palmer* who had never crossed the Antarctic Circle, including everyone in the crab team except for Maggie and Sven, had now received fair warning. The gala kicked off to a raucous start when King Neptune, draped in long robes and wearing a crown, was joined by the ship's crew dressed as pirates, several in drag. The ship's captain attended the gala, but with appropriate attention to decorum, he watched the activities from the sidelines. As is the custom, the inductees arrived ready to perform and wore their underwear on the outside of their clothing. Poetry, storytelling, songs, and skits marked the occasion, ingeniously poking fun at life aboard ship. Then it was time for the secret ceremonial induction. Roberta revealed only that blindfolds and buckets of seawater were involved.

The crab team spent the next several weeks aboard the *Palmer* sampling plankton for traces of crab larvae and searching for king crabs among the thousands of digital images recorded from Marguerite Bay. Now in the Bellingshausen Sea, the ship was tasked to retrieve several oceanographic subsurface data buoys that another research vessel had previously deployed. Unfortunately, only one of the buoys responded to the electronic release signal and the chief scientist and ship's captain deemed the rest irretrievable after repeated attempts to snag them with a grappling hook. As planned, a month into the cruise, the *Palmer* rendezvoused with the *Oden* to transfer the crab team and their submersible over to the icebreaker for the balance of the trip. To accomplish this task the *Oden* tethered itself with large ropes to the edge of a large sheet of sea ice. The *Palmer* then pulled up alongside the *Oden* and the ship's crew placed a gangplank between the two. The entire crab team, with the exception of the two Danish graduate students who stayed on the *Palmer*, transferred to the Swedish vessel. Once everyone was aboard, a crane transported the luggage and laboratory supplies, and the ships parted ways. The *Palmer* departed to sample the polynya in the Amundsen Sea while the *Oden* continued nearby with its own research objectives.

Maggie and Roberta reported that life aboard the *Oden* had its cultural bonuses. Unlike the *Palmer*, which operated under strict naval regulations as a dry ship, alcohol was permitted on the *Oden*. Scientists and crew could gather in the bar after a shift or on special occasions to socialize and have a glass of wine, beer, or the popular Swedish schnapps. The cabins on the *Oden* were larger than those on the *Palmer*, and the water in the showers lacked the strong organic odor of the Antarctic planktonic marine alga *Phaeocystis* that plagued the smaller ship's showers. The Swedish cooking was good meat-and-potatoes fare, and on special occasions the cook served pickled herring and the traditional Swedish *lutefisk*. With a little cookbook sleuthing, I learned cooks prepare this delicacy by soaking dried cod in water for five days and then for another two days in water with lye. This process causes the fish to swell and lose protein, giving it a jellylike consistency. At this point, the fish is caustic and inedible. The trick to making it palatable is to soak the fish in lye-laced water for five more days. Twelve days into the recipe the fish is ready to be steamed or parboiled. Having never indulged in lutefisk, I must refrain from up-front judgments. However, despite a notoriously offensive odor, I was more concerned by the caveat in the cookbook that stated: "It is important to clean the lutefisk and its residue off pans, plates, and utensils immediately. Lutefisk left overnight becomes nearly impossible to remove. Sterling silver will be permanently ruined." The way I figured it, if sterling silver was endangered, this can't bode well for one's digestive organs.

As the crab team analyzed the thousands of images of the Antarctic slope captured by the submersible, the initial counts of king crabs rapidly accelerated from the tens to the hundreds, which signaled that the species is on a collision course with archaic marine communities of the shallow Antarctic shelf. Weakly shelled snails and clams and sluggish brittle stars and starfish would be no match for this army of clawed predators. These rich marine communities contain marine organisms that have been free of crushing predators for millions of years.

*The Paleozoic-like antiquity of the organisms* that comprise the shelf communities has its roots in the geological origin of Antarctica. Two hundred million years ago—in the early portion of the geological epoch known as the Jurassic period—a future Antarctica was but a small part of a vast supercontinent known as Gondwana. This amalgamation of continents included most of the major landmasses found in the present-day Southern Hemisphere—Africa, Australia, South America, and portions of New Zealand, Madagascar, and the Indian subcontinent. About 170 million years ago, through the action of plate tectonics—the movement of the plates of the earth's outer rocky crust—the eastern section of Gondwana, made up of Madagascar, India, Australia, and a future Antarctica, drifted off to the west of Africa, opening up the South Atlantic Ocean. India headed north across the equator. Finally, somewhere between thirty to forty million years ago, Australia and South America pulled away from Antarctica, leaving it, as today, isolated in its polar position. The separation of what is now Chile and Argentina from the continent of West Antarctica opened up the Drake Passage, joining the Pacific and Atlantic Oceans. The land barrier that had forced cold waters of the Southern Ocean to the north to be exchanged with warmer tropical waters no longer existed. Like a massive dam bursting, the Southern Ocean's icy waters pushed through the Drake Passage, catalyzing the world's largest current, the Antarctic Circumpolar Current, or ACC. The formation of the ACC had the effect of slamming the door of a massive deep freezer, sealing the continent's frigid fate. Today, what constitutes 80 percent of the world's fresh water is locked up in a two-mile-thick mantle of ice draped over a continent the size of India and China combined. Despite some limited interchange of marine life between Antarctica and South America that continues to this day, the marine organisms of Antarctica found themselves isolated by the icy ACC from the rest of the Southern Hemisphere. Antarctic marine species had little choice. They either adapted to the frigid temperatures, or they went extinct.

Seymour Island is one in a chain of sixteen islands located off the northernmost tip of the western Antarctica Peninsula. It is hilly, rocky, and windswept. Argentina maintains a year-round station on the island—Marambio Base—one of the few stations along the Antarctic Peninsula equipped with a runway for wheeled aircraft. The island is well-known among paleontologists because it is the best place in Antarctica to collect marine macrofossils. Its rich fossil beds first came to light in 1882 when a Norwegian ship captain, Carl Anton Larsen, discovered a bounty of long-extinct species fossilized in its rocks. A century later, the discovery of a marsupial fossil on Seymour Island provided the first proof that land mammals had once roamed Antarctica. Rich Aronson has spent endless hours hiking up and down the steep slopes of Seymour Island, digging through the loose dirt and carefully wrapping each fossil he finds before securing it in his backpack.

A fellow scientist with whom I often discuss climate change is E. O. Wilson, the world's authority on ants. We have enjoyed a good laugh that the one major continent on the planet that has no ants is "Ant"arctica. So you can imagine my delight when, during a field season at Palmer Station, I discovered a column of tiny ants marching across my desk. They had probably arrived with our recent shipment of fresh fruit from Chile, and, not surprisingly, they disappeared the next day. But their brief appearance serves as a reminder that as climate change continues to accelerate in Antarctica, invasive species, whether coming from deep water or introduced by research and tourist ships, are sure to establish themselves with greater frequency. To date, scientists have documented few examples of introduced marine species. The only three cases on the Antarctic Peninsula include a temperate brown algal endophyte (an alga living inside another alga) that our team discovered near Palmer Station; a temperate green alga that has established a population in the intertidal on Half Moon Island; and an Arctic spider crab that Brazilian polar marine scientists dredged up from the seafloor off the northern tip of the peninsula. The spider crab was collected on a single occasion and probably had no mate, but the fact that an Arctic crab or its larvae survived what must have been a challenging transit across the equator in the ballast water of a ship, to later be dredged up alive in the waters of

Antarctica, speaks volumes about the growing potential for the introduc-
tion of invasive species.

Rich Aronson's first trip to Seymour Island was in December 1994.
He and his colleague, Dan Blake, an expert of echinoderm paleontology,
had arrived via a C-130 cargo plane flown by the New York Air National
Guard. As guests of the Argentine Air Force, Rich and Dan enjoyed free
housing and meals at Marambio Base for a little over a month. Their goal
was to evaluate whether the dense populations of brittle star and crinoid
fossils in the Eocene La Meseta rock formation told a story of how rare
fish predation was in ancient Antarctic seas. By examining hundreds of
fossils, they were able to demonstrate that the incidence of arm damage
and regeneration caused by predators was almost nonexistent, just as
they had hypothesized. Rich and Dan returned to the island in December
2000, this time setting up a field camp to collect fossilized shells of clams.
They ran isotopic dating tests on the clam shells to determine what the
sea's temperatures had been over the past forty million years.

By determining the types of fossilized invertebrates present and
knowing the age of the rocks from which they were found, they concluded
that the Antarctic seafloor had been free of crushing predators ever since
the advent of the Drake Passage had cooled down Antarctic seas. At that
time, fish with crushing jaws and crabs with crushing claws began to disappear. The marine invertebrate species that had subsequently thrived
belonged to a group of organisms known as filter feeders (brittle stars,
crinoids, sponges, soft corals, clams, brachiopods, or hydroids). For all
practical purposes, present-day shallow Antarctic seafloor communities
were similar to the archaic seafloor communities of the Paleozoic era
some 542 million to 251 million years ago. As I mentioned in chapter 5, these fossilized marine invertebrates were generally very poorly
calcified just as their living representatives are today. Their shells are
thin and fragile—I have picked up clams and snails in Antarctica and

crushed them easily in my hand. As such, they are extremely vulnerable to both ocean acidification and crushing predators. Rich, independent of Sven, similarly predicted that rising sea temperatures could allow the reintroduction of crushing predators, such as king crabs, into Antarctic waters over the next fifty years.

The living accommodations for Rich and Dan's stay at the Argentine station on Seymour Island in 1994 were modest but downright elegant compared to the field camp they occupied in 2000. This primitive camp consisted of two mountain tents and a bucket fitted with a toilet seat. Rich quipped to me later that "sitting on the toilet out in the open with the wind howling around us was delightful." Because their stay in the field camp was a relatively short four days, the *Gould* anchored nearby. The ship's cook insisted on bringing Rich and Dan hot dinners delivered each evening by zodiac. On the day of their departure from the island, they threw themselves with abandon into the showers to remove the layers of grime that only paleontologists digging in the dirt could accumulate. That evening, as he climbed into his bunk in the chief scientist cabin, Rich was filled with regret and melancholy knowing that this would likely be his last opportunity to ever see his beloved Seymour Island. But he need not have worried. When he woke the next morning, Seymour Island was still in view. In fact, the ship was closer to the island than when he had gone to bed. During the night, the sea ice had closed in around the *Gould* and, much like Shackleton's *Endurance*, they had been besieged in the pack ice. They remained stuck fast in the ice for three full days before the *Gould* finally broke free.

⌒

*By early January 2011*, the *Oden* and the *Palmer* reached the Ross Sea. In contrast to their time aboard the *Palmer*, the crab team had fewer opportunities to lower the sled from the *Oden* to the seafloor to search for crabs. Unable to pull the sled at the appropriate speed, the *Oden* on several occasions tied off to fast ice so that the sled could be lowered. The limited surveying that could be accomplished had revealed at least one king crab, indicating that crabs were also present on the slope of the Bellingshausen and Amundsen Seas.

When the *Oden* arrived in McMurdo Sound and began to break the ice channel for McMurdo Station, life on board changed. To break through the six-foot layer of sea ice, the ship would literally power itself up onto the ice like a huge elephant seal. Warm water, heated by the ship's engines, sprayed from the bow jets over the surface of the sea ice to help weaken it. Ultimately, the sheer weight of the front end of the ship would break through the ice. The ship would then back up and repeat the process. Roberta recounted that the noise on board the ship while it broke through the ice was pervasive, and it reminded her of "a cell phone set on vibrate, buzzing nonstop, or pots and pans shaking about like in an earthquake." Some of the crab team had become accustomed to the noise and vibrations; others never did.

Over the month spent on board the *Oden*, Maggie and Steph completed a meticulous evaluation of the images of the Marguerite Bay seafloor. They counted astonishing amounts of crabs, numbering in the hundreds. If one assumed that these counts were indicative of average crab densities representing the seafloor environments sampled by the submersible's camera, then the numbers climbing the Antarctic slope of Marguerite Bay alone would measure in the millions. How large an ecological impact would king crabs have on the shelf seafloor? While the precise outcome is difficult to predict, the impact would be dramatic. If king crabs move up and onto the shelf in large localized groups, in the first year or two, like a tornado, they could be expected to decimate localized populations of vulnerable shelf invertebrates as they encounter them. Over several decades, should the crabs become firmly established on the shelf, the ecological consequences would likely entail fundamental shifts in community makeup, with some species resisting crab attacks and others vanishing. As scientific studies of Antarctic marine seaweeds and marine invertebrates have shown, this unique ancient marine biota represents a priceless repository of chemical compounds that could fight cancer, AIDS, and influenza, as well as produce novel antibiotics.[2] Species eradication could come at a terrible price. A fishery might control future populations of king crabs on the Antarctic shelf and is worth consideration. However, crab fishing in treacherous Antarctic waters would be risky. Much like in the north Pacific, Antarctic storms can rapidly generate huge waves and chaotic seas, making the recovery of crab pots dangerous. Despite the challenges, I already bear evidence of interest: an Arctic crab fisherman emailed me and asked when he might head his crabbing vessel south.

With an invasion of king crabs looming, the potential impacts on shelf communities have not been lost on the media. As the *Oden* completed the final days of its voyage, I briefly visited Palmer Station while leading a Climate Change Challenge philanthropic cruise to the Antarctic Peninsula for the travel company Abercrombie and Kent. Three international news reporters who were visiting the station for several weeks interviewed me about the king crab invasion. When the *Oden* docked at McMurdo Station, news reporters interviewed Sven there. The interviews at Palmer and McMurdo resulted in articles in the *Economist*, *Discovery News*, and the *Washington Post*, and former Vice President Al Gore's office requested a photo so that he could include the king crab invasion story in his Global Climate Change presentation. Rich's off-the-cuff remark quoted in the *Discovery News* article summed it up nicely: "As the surface waters warm up, that will make it possible for the crabs to come running over the top and raise hell with the bottom communities."[3]

I was made aware of additional evidence of the pending Antarctic crab invasion via a phone call in mid-September 2011 from a reporter with the *London Times*. He asked if I would be willing to preview an article on king crabs invading Antarctica in the prestigious British science journal *Proceedings of the Royal Society*[4] to be released the next day. I had heard rumors of this exciting discovery and accepted the offer. When I downloaded the early-release article, I learned that a research team led by Craig Smith, a biological oceanographer at the University of Hawaii, had happened upon a massive population of king crabs in March 2010 during a routine remote survey of the Palmer Deep, an underwater canyon gouged into the shelf of the Antarctic Peninsula just ten miles off the coast of Palmer Station on Anvers Island. Craig and his team were aboard the *Nathaniel Palmer*, and had been stunned when their remote submersible transmitted digital images revealing bright red spiny king crabs, some measuring up to three feet from arm tip to arm tip. A return cruise to the Palmer Deep less than a year later allowed Craig and his colleagues an opportunity to collect live crabs using a trawl towed behind the *Laurence Gould*. Some of the bigger female king crabs were carrying large masses of ripe eggs—an indication that the population in the canyon was probably self-sustaining. Despite the crabs in the Palmer Deep living at lower depths than those on the shelf, the fact that the crabs may have crawled seventy-five miles across the shelf from deep water to populate a canyon on the very edge of the continent, or, alternately, that crab larvae had survived a swim across the shelf to settle into the canyon, is yet another ominous sign that Antarctic near-shore waters are poised for a crushing invasion. Also ominous is the fact that the canyon is so close to land that on a clear day one can see a ship positioned over it from the top of the glacier behind Palmer Station. Based on the canyon's dimensions and the submersible crab counts, Craig and his colleagues estimate that the total crab population is an impressive 1.55 million individuals.

# Chapter 7

# Ghost Rookeries

## THE DECLINE OF THE ADÉLIE PENGUIN

*E*xhausted after her long swim, the lone Adélie penguin pushed and prodded her way through the bits and pieces of loose floating ice, thrusting herself up onto the slippery intertidal rocks of Torgersen Island, a tiny outcrop off the central western Antarctic Peninsula. She clambered to her feet, her natal instincts guiding her to the eastern reach of the island. Her path was well worn, garnered by some thirty generations of forebears who on this very week each spring scrambled across this same stretch of the island. She paused, listening to the braying of those that had already arrived. The trumpeting calls of nesting mates from recent years were absent; nonetheless, she directed her attention to constructing a nest out of small, smooth pebbles. Despite the dwindling numbers of potential mates, she went about her business, finally approving the courting head nods, flapping displays, and guttural brays of a persistent male. Soon, two eggs sat warmly below her fine dense feathers. The embryos developed rapidly within their shells, their genomic software orchestrating the hardware necessary to communicate, to move through icy waters like a porpoise, and to forage on krill that provide the crux of sustenance for so many Antarctic marine animals.

The wind calmed beneath the overcast sky under which the first snowflake gently fluttered earthward on this late-spring day. Knowing instinctually that her chicks were close to hatching, the Adélie vigilantly covered her eggs as the snowfall began to accumulate. At first, only a thin white veil covered the rocky terrain, and later, as the evening light diminished, the snow formed a dense carpet over the entire nesting ground. Dawn revealed a haunting scene on this miniscule island: the penguin still nested, although she was completely buried in the snow, with only a small opening left from which to breathe. Sadly, her two unborn chicks would eventually, inevitably, drown in the meltwater of this unseasonable snowstorm, a storm fueled by an increasingly moist and warming climate. Each fated chick thus became a metaphor for the potentially cataclysmic juxtaposition of climate-related events now gathering force over the Antarctic Peninsula.

She was hardly alone. The Antarctic Peninsula is home to a remarkably rich abundance of seabirds and marine mammals. Skuas, terns, petrels, and blue-eyed shags nest along its rocky shores. Chinstrap and gentoo penguins join Adélies as they swim through its icy emerald seas. Minke, sei, fin, right, and humpback whales, along with the occasional blue, break the surface to spout once or twice before a parting stroke of the fluke. Killer whales, which are actually large dolphins, point their heads straight out of the water in "spy hops," scanning the ice edge for their penguin prey. Crabeater seals, which currently hold status as the most abundant seal on earth, lounge about on ice floes. Female Weddell seals, with their unique ice-chipping teeth, lay about on the annual sea ice next to tooth-chipped ice holes with their charismatic pups. Elephant and fur seals abound, while powerful leopard seals prowl the near-shore waters, feeding on krill, penguins, and small elephant and fur seals. Although each of these animals has its passionate champions, penguins are the most beloved Antarctic animals in popular culture, thanks in large part to blockbuster films like *March of the Penguins* and *Happy Feet*. Each film propagates the image of the penguin as an amusing, lovable, and nonaggressive animal.

Adélie penguins certainly conform to this stereotype with their feather-tight black-and-white tuxedo look and Chaplin-esque demeanor. I witnessed their remarkable nest-building behavior for the first time during a visit to the Adélie rookery at Cape Royds in 1989—named for Lt. Charles W. R. Royds of the Royal Navy and the 1901–1904 British

Discovery Expedition. About fifteen miles north of McMurdo Station on Ross Island, approximately four thousand breeding pairs had constructed circular nests from small walnut-sized stones. Resembling thousands of miniature campfire rings, the nests dotted the hillsides and valleys of the rookery. In each, a female typically deposits two eggs. I sat on a rocky outcrop and watched for hours, transfixed. I laughed out loud when I realized that Adélies periodically vacate their nests to steal a stone from another, not realizing that during their brief absence a neighboring penguin has simultaneously pilfered a stone from their own nest. Their behavior was the very essence of a zero-sum game.

Eleven years later, in 2000, during my first research expedition to Palmer Station, I jumped into a rubber zodiac boat to visit the Adélie penguin rookeries on surrounding Torgersen, Humble, Christine, and Cormorant Islands. After cutting off my zodiac's engine and securing the boat's bow line to the eyebolt driven into a rock at the island's designated landing site, I deployed a stern anchor to keep the zodiac from swinging around. The raucous cacophony of nesting penguins greeted me. The Adélie colony on Torgersen Island is by far the largest and the most impressive in the Palmer region. Paleontological evidence derived from radiocarbon dating of excavated penguin bones indicates that for at least the past seven hundred years, Adélie penguins have successfully nested on Torgersen Island. The dry ground is a nesting haven since Adélie eggs are incapable of surviving exposure to ice and snow. Skuas, large hawk-like birds with powerful talons that can weigh as much as three and a half pounds and have a wingspan of up to four feet, constantly cruised the rookery, watching for the slightest opportunity to swoop in on an exposed egg. Later in the summer nesting season (January and February), young chicks left on their own may become prey while their parents forage for krill, silverfish, or squid to alleviate their state of near starvation during egg incubation and to nourish their rapidly growing, ravenous, and noisy offspring. Often, skuas scare penguins off their nests just long enough to swoop in and grab one of the two eggs. Once, a seemingly abandoned chick was carried off, screeching, right before my eyes. By early fall (March) the Adélie chicks molt their soft, gray, downy feathers, revealing their ubiquitous tuxedo-esque layer of densely packed, insulatory, adult feathers. Soon thereafter, they take to the sea along with their departing parents.

Dr. Bill Fraser, president and lead investigator of Polar Research Oceans Group, a corporation located in the appropriately cold-temperate town of Sheridan, Montana, is in his mid-fifties and has a soft-spoken demeanor, thinning blond hair, and the weather-beaten skin of a charter fishing vessel captain. He has worked with the Adélie penguins on Torgersen Island and a cluster of neighboring islands for a remarkable thirty-five years. Bill's research on the Antarctic Peninsula, funded by the National Science Foundation, has contributed to his status as one of the world's leading authorities on the population biology and ecology of Adélie penguins. Over the years, he has increasingly suspected that rapid climate change bringing warming air temperatures to the western peninsula would also bring about greater levels of snowfall as well as a prolonged snow season. He was right on both accounts. As the Antarctic peninsular air has warmed, its sub-freezing air has become increasingly humid, facilitating additional snowfall.[1] Fortunately, scientists at the Ukrainian Station, Vernadsky, just thirty miles south of Palmer Station, have long recorded midwinter air temperatures. Their winter measurements show a clear pattern of increase by a total of a little over 11 degrees Fahrenheit over the past 60 years, meaning that midwinter air temperatures have gone up almost 2 degrees per decade. Imagine living somewhere with typical winter temperatures in the mid- to upper twenties suddenly warming to an average well into the thirties. Where you used to shovel the snow off your driveway week after winter week, you'll now find it difficult to even make a snowman.

Now, for the first time in Fraser's Antarctic career, an unseasonable blizzard had buried the female Adélies who had arrived to lay their eggs in early November. Because Adélies are smaller than their cousins, the king and emperor penguins, only the tops of their heads were visible in some cases, and only when one peered down soccer ball-sized holes in the snow. They looked forlorn, even pathetic, each as obscured as a golf ball buried deep in the rough. Tragically, the eggs were incapable of withstanding the brutal exposure to ice and meltwater and the embryos died. The next generation of Torgersen penguins, who had successfully nested on the island for seven hundred years, was wiped out.

Unseasonable snowstorms are only one consequence of the climate change that has decimated this Adélie penguin population. Donna Patterson, an Antarctic seabird researcher and Bill Fraser's wife, has conducted studies detecting differences in penguin adaptation between study sites based on whether a given site is facing to the north or south and whether the populations are small or large in number. These factors are linked directly and indirectly to climate change. Directly, penguin groups in landscapes with southern exposures are subject to increased levels of snow accumulation that drowns eggs during snowmelt. Indirectly, with fewer chicks to replace adults in the population, smaller numbers of adult individuals are available to help ward off aggressive skuas that prey on eggs and chicks.

The Adélies face additional challenges. Warming sea and air temperatures along the north central Antarctic Peninsula have reduced both the distance the pack ice extends offshore and its duration during the year by almost half over the past three decades. Dramatic reductions in the winter pack ice offset the delicate balance between the penguins' intake of food resources and necessary energetic expenditures. The ice provides a bridge for the Adélies to cross to reach rich krill grounds as well as a platform from which to forage. As the pack ice continues to recede as the climate warms and as the krill continue to decline significantly along the northern half of the peninsula, Adélies will likely have to swim farther and farther offshore to reach feeding grounds, expending critical energy that can make the difference between a successful breeding season and mass untimely death. These factors help explain why Bill and his research team have documented the loss of approximately 12,000 of the 15,000 Adélie breeding pairs, approximately 80 percent, that occupied this region of the peninsula as little as 35 years ago. Fraser predicts that Adélies will soon be entirely gone from the small islands that surround Palmer Station.

The ecological consequences of the loss of the annual pack ice along the peninsula are brought into sharp focus when one considers the direct impact on Antarctic krill. Despite their relatively slow growth rates, the swarming populations can attain dimensions that are simply mind boggling, in some cases measuring on scales extending for hundreds of square kilometers. Only fairly recently have krill researchers discovered that their life history is intimately tied to the ebb and flow of the annual pack ice. Here, as juveniles, they spend their winters living among the three-dimensional latticework of the undersurface of the ice, which forms a refuge. Small microalgae, or diatoms, grow in greenish mats on the undersurface of the seasonal pack ice, providing juvenile krill with a critical source of nutrition. Quite simply, as the pack ice goes away, so do the juvenile krill.

In striking contrast to the threatened Adélie penguins along the Antarctic Peninsula, other brushtail penguins (named for their distinctively long tails)—including both the gentoo and chinstrap—have relatively recently established breeding colonies along the central western peninsula (chinstraps first in 1975, then gentoos in 1992). Both species are quite easy to identify. Gentoos have a bold white stripe that looks a bit like a bonnet perched on their heads. When fully grown, they stand about three feet high, slightly taller than their Adélie and chinstrap cousins who measure about two and a half feet in height. The only penguins larger than the gentoo are king penguins and, of course, the majestic emperor penguins. Gentoos are very adept in the water and capable of diving to some two hundred feet. They repeatedly burst from the sea surface and then dive back down with strong flips of their wings. They are the fastest of penguins, having been clocked swimming underwater at up to twenty-two miles per hour. Gentoos build the largest nest of the three brushtail penguins. Scientists have recorded gentoos gathering up to 1,700 small smooth stones to construct a circular nest that is about two feet in diameter. They ascribe a special significance to their nest stones, with males known to present an offering of a single meticulously selected stone to a female to seduce her into courtship, despite an ultimate lack of fidelity over their lifespan. Chinstrap penguins, named for the distinct black band under their snow-white chins that makes them appear comically "helmeted," are bold and aggressive. Similar to the Adélie and gentoo, chinstraps build their nests out of small stones.

Both gentoos and chinstrap penguins had previously been primarily confined to the northern reaches of the peninsula and to the subantarctic, including the magnificent South Georgia Island where Adélies occur in small numbers. King penguins occur there in truly spectacular abundance with some half a million breeding pairs present. Gentoos are also abundant with 105,000 breeding pairs, while chinstraps number around 6,000 breeding pairs. Scientists on South Georgia have found that with so many penguins sharing common ground, they segregate from one another on the basis of their breeding cycles, foraging behaviors, and life-history traits. As of 2009, gentoo breeding pairs in the Palmer region numbered about 2,400, and chinstraps about 310. Provided they can secure sufficient krill resources, these penguin populations are likely to continue to grow because their life histories, in contrast to the Adélies', are not intrinsically linked to the ebb and flow of the annual pack ice. For example, they do not depend on the annual sea ice as a platform to access rich offshore krill populations. Moreover, the seasonal timing of egg laying is such that gentoos and chinstraps nest in the Palmer region about three weeks later in the austral summer than the Adélies, when snowstorms have ceased, the snow has melted, and when little threatens their offspring. My sad, but scientific, prediction is that these penguin populations may soon outnumber the Adélies from the central to the northern tip of the peninsula (presently, they are about equal in number), a prediction that parallels a general pattern of climate-induced species replacement as global warming alters the climate. Some may argue that this phenomenon is nothing new, but such species shifts have historically occurred over millennia rather than over a few short decades.

The encroachment of nesting gentoos is becoming increasingly evident. During a Climate Change Challenge Mission in 2009 on the cruise ship *Minerva*, Patricia Silva-Rodriguez, a Uruguayan ornithologist, and I discovered approximately seventy-five gentoos happily nesting within the confines of a long-established Adélie rookery at Brown Bluff on the eastern side of the Antarctic Peninsula. This encroachment of gentoos was nothing new, as they are well-known to take over nest sites of other penguin species within rookeries, probably as a mechanism to reduce predation by skuas. However, our discovery indicated that nesting gentoos in the Adélie rookery are solidifying their colonization of the eastern Antarctic Peninsula as the sea ice retreats and ice sheets increasingly break apart in this region.[2] Rapid climate change is causing unprecedented southerly shifts in nesting gentoo penguins, which is indicative of future changes in the ecological structure and function of the northeastern Antarctic Peninsula. In other words, the remarkable change scientists are seeing in the ecology of penguins on the western Antarctic Peninsula is likely to soon follow suit on its eastern flanks.

In any given field season, when our research team first arrives at Palmer Station, usually in the early austral fall (February), we can encounter gentoo, chinstrap, Adélie, and even, quite rarely, the odd, wayward king penguin standing near the seawater intake pipe that plumbs our research aquarium facility. Some remain for a day or two. These visits may be motivated by the need to avoid a neighboring leopard seal, or perhaps they are simply driven by curiosity. Both species, like the Adélie, typically lay two eggs, and after their parents wean them, the young join *crèches,* or clusters of juveniles, ensuring greater safety in their numbers. Gentoos and chinstraps feed primarily near the shore on krill; they also include amphipods, squid, and fish in their diets. Scientists are uncertain whether the significant krill depletion that has taken place along the central and northern peninsula will eventually take its toll on these penguins or whether they will adapt to alternative prey or move to regions richer in krill. Bill Fraser and his team conducted some banding studies with Adélies near Palmer Station in the mid-1990s and found that very few of the birds they provisioned with wing bands were observed elsewhere along the peninsula. This research is not good news: it indicates that the Adélies may not have the inherent flexibility to pack up and move south as the climate warms. Those that are disappearing from Palmer region are likely doing just that: disappearing.

In contrast to the highly visible Adélies waddling along the shores, the powerfully built and highly predatory brown skuas common along the western Antarctic Peninsula are harder to spot. Brown skuas nest upon rocky, open terrain, and their mottled earthy-brown coloration allows them to blend in with the landscape, rendering them difficult to detect when standing on the ground or sitting on the shallow earthen depressions that mark their nests. They return each year to the same nest site and are generally faithful to a lifelong mate. The predatory, dive-bombing skuas are not to be taken lightly. One of these feathered rockets knocked an Antarctic biologist unconscious. The impact of a skua dive-bombing someone's skull would be equivalent to an NFL quarterback braining a bystander at close range with an errant pass.

Besides being downright aggressive, skuas are also kleptoparasites, an awkward term referring to their nasty habit of forcing other seabirds to drop whatever they're feeding on at the moment so the skuas can steal and eat it. Moreover, adult brown skuas are quite capable of capturing and consuming other seabirds. And, of course, their true claim to infamy along the peninsula and in other regions throughout coastal Antarctica comes from their egregious habit of raiding penguin rookeries. With the dramatic reduction in the numbers of Adélie penguins living along the peninsula—and despite some replacement by increasing numbers of gentoos and chinstraps—an indirect impact of climate change in this region may include skuas having to shift their diet to alternative prey, compensating for a reduction in access to penguin eggs and chicks.

Out on Torgersen Island's Adélie rookery, I have patiently watched the rhythm of the hunt. Perched on the rocks at strategic intervals, lone skuas stand watch. Any event can crack the door for a hungry skua: a squabble between two adult Adélies, a distraction caused by a rummaging sheathbill, a nest vacated for just a moment to collect a nest stone. They are opportunists extraordinaire. Skuas can pick the eggs out of nests in the blink of an eye and fly off with squawking young clutched in their beaks. Perhaps the saddest instance is when a skua drives a helpless chick away from the nest and then pecks it to death. In the vast majority of circumstances, however, parent Adélies defiantly defend their eggs and chicks from the marauders.

Over several field seasons at Palmer Station, I had the good fortune of becoming friends with Donna Patterson, who is more commonly known as "Patterdo." She acquired this nickname when someone on the IT staff at Antarctic Support Associates, one of the companies that has in the past provided logistical support for the U.S. Antarctic Program, assigned email addresses by taking the first six letters of one's last name and combining it with the first two letters of their first name. Patterdo and her birding assistants intimately introduced me to another large, aggressive, feathered predator that nests along the western Antarctic Peninsula, the giant petrel (also known as the southern giant petrel). Climate warming seems to have been kinder to the petrels in the Palmer region than it has to Adélies, in part because their foraging flights take them far afield, decoupling them from local food resources. One of the petrel's primary nesting sites is near Palmer Station on nearby Humble Island, where Patterdo's team has been monitoring the population since the mid-1970s.

We made an investigative field trip to the island one early fall day at a time when the birds were raising their chicks. Donna explained to me that these giant petrels are extremely opportunistic in their feeding habits. They prey upon the usual assortment of krill, fish, and squid, *and* they also exploit the chicks of other seabirds by either drowning them or using their wings to beat them to death. They also eagerly consume the carcasses of penguins and seals, and they have thus deservedly appropriated the common name "stinkers." Whalers labeled them "gluttons." The birds are massive when full grown, weighing approximately eleven pounds, measuring three feet in length and boasting an impressive wingspan of six feet.

I watched as Patterdo and her birders worked carefully to weigh each scruffy grayish-white chick (not nearly as endearing as penguin chicks) as they, and their protective parents, can open up a gash with their hefty beaks. These feisty chicks have the raw muscle power of a full-grown hawk. This aggressive behavior is apparently a common enough occurrence that a scientific publication entitled "Injuries to Avian Researchers at Palmer Station, Antarctica from Penguins, Giant Petrels, and Skuas" exists.[3] Patterdo's team must spend a considerable amount of time habituating the giant petrels to humans before handling the birds.

Unlike the plummeting numbers of Adélies on Torgersen and neighboring islands, giant petrels nesting on Humble Island have a thriving population that has doubled over the past thirty years. Each year the females, in contrast to Adélie and skua females, lay only a single egg in the late spring or early summer and then incubate it for a period of approximately two months. Once the chicks hatch, the mothers hold them in a feathered brood pouch. Even though the females provide the brood pouch, giant petrels mate for life, unlike the vast majority of birds, some of which form pair bonds that last only for several years. The chicks, awkward in their movements, eventually learn to fly and depart the colony after about four months. Patterdo and her team have equipped some of the giant petrels with radio transmitters and have used satellite tracking techniques to document the birds' extensive forays in search of food. Some foraging flights literally span thousands of miles.

The population on Humble Island may be doing so well because these birds do not appear to forage in regions of the Southern Ocean subject to intensive long-line fishing operations. Unfortunately, giant petrels in other regions of the Southern Ocean are not as fortunate and continue to suffer high mortality from getting hooked on the baited lines deployed behind the fishing vessels. As many as four thousand of these magnificent birds are hooked each year, and, sadly, they drown as the fishing nets descend into the sea. The impact of long-line fishing on this species has now become so severe that the Scientific Committee on Antarctic

Research (SCAR) recently developed a proposal to list the southern giant petrel as a "Specially Protected Species" in the Antarctic Treaty. Moreover, both this and the northern giant petrel are now classified as "Near Threatened" by the International Union for Conservation of Nature and Natural Resources. So far, the negative impacts of summer blizzards experienced by the Adélie penguin do not appear to be taking a similar toll on the Humble Island giant petrel population. However, in 2003, scientists found several giant petrels starved to death and literally frozen to their unhatched eggs on the Frazier Islands, located near Australia's Casey Base on the Antarctic coast about two thousand miles due south of Perth. Australian scientists surmise that the birds were likely caught in a late blizzard and died while incubating their eggs. Thus, giant petrels nesting along the Antarctic Peninsula could be vulnerable to snow-related mortality, particularly given that climate warming models predict a continued pattern of increased and more persistent snowfall.

In sync with recent climate changes and in step with the gentoos and the chinstraps, some subantarctic seals have begun to colonize the Antarctic peninsular region. For example, in the past thirty years, the southern elephant seal, which derives its name from its elongated elephant-like proboscis and historically lives on the subantarctic islands, has extended its range south along the peninsula as air temperatures have warmed. In the span of but a single decade, I have personally witnessed increased numbers of these seals in the vicinity of Palmer Station. Indeed, not long ago, I opened the front door of the biolab early one morning, caught a whiff of something disturbingly offensive, and then almost tripped head over heels on a massive bull elephant seal that was sound asleep below our station doorstep.

For centuries, sealers hunted these animals, particularly the colony on remote Kerguelen in the southern Indian Ocean, now home to the Port-aux-Français research station. The hunters struck the animals over the head with clubs, then flanked them and hauled their blubber off by ship to nearby Port Jeanne D'Arc, about thirty miles southwest of Port-aux-Français station, the site of a Norwegian whaling and sealing industry founded in 1908. Here, seal processors rendered their fat into fuel oil and shipped it north. And there is a lot of blubber to be taken from a single seal. Full-grown male bull elephant seals can be truly massive: the largest recorded male measured twenty-two and a half feet and weighed eleven thousand pounds. Females are typically about half the length and a third of the weight of males. Elephant seals, which can appear slug-like in their torpor, are quite capable of moving faster than we humans can sprint for short distances over terra firma. Fittingly, the Port-aux-Français research station is also home to the world's one and only elephant seal crossing sign. A large yellow traffic sign bearing a diagram of an elephant seal sits alongside a small station road that separates the research station from the Gulf of Morbihan. To this day, I believe the warning sign was the result of a rather unfortunate encounter between an elephant seal and a vehicle driven by a scientist who had consumed one too many glasses of Bordeaux.

As a college student at the University of California–Santa Cruz in the mid-1970s, I had the unique opportunity to conduct undergraduate research on the diving behavior of northern elephant seals with Dan Costa, then a budding doctoral graduate student and now a highly respected marine mammalogist. Dan and I visited the elephant seal colony at the Año Nuevo Preserve north of town, and I helped him first narcotize several adult seals, and then glue time-depth recorders to their backs. These pioneering diving studies revealed that elephant seals forage at amazing depths, ranging from 900 to 1,800 feet. More recent studies have shown they are capable of diving 4,500 feet and can stay submerged for as long as 80 minutes. The physiological adaptations necessary for an air-breathing marine mammal to accomplish this feat boggles the scientific mind. Such deep dives allow elephant seals to capture and consume skates, rays, squid, octopus, and bottom fish. Recently, I spoke with Dan at a science meeting and learned that he is now attaching satellite radio transmitters on elephant seals and exploiting their deep diving capabilities to measure key oceanographic features such as temperature, salinity, and depth of the Southern Ocean. Elephant seals are providing significant oceanographic environmental data relevant to climate change research that even the most sophisticated equipment on oceanographic ships cannot match.

In 1955, scientists first documented elephant seals in the region of Palmer Station. But it wasn't until almost thirty years later, in 1983, that scientists first reported that elephant seals had established breeding colonies that resulted in successful reproduction. Since that time, they have established additional successful colonies, especially in the region known as Elephant Rocks, a group of prominent rocks located between the northwest entrance of Arthur Harbor and Torgersen Island. During the mating season, the young males practice fighting with one another, and older males challenge the dominant bull for mating rights with females in the harem. When the weather conditions are calm enough, their guttural roars, visceral grunts, and the slapping of flesh in combat echo off the face of the Marr Glacier. While their recent migration into the Palmer region may at first seem a positive impact of climate change, Donna Patterson is concerned that the invasive elephant seals are now competing for space on islands with colonies of giant petrels, and thus the growing seal population may eventually have a negative impact on the success of the petrel colonies.

❧

*In March 2010,* a film crew producing a two-part series for the televised travel documentary *Globe Trekker* visited us at Palmer Station. Even though it is perhaps among the most successful travel shows and now seen in some forty countries over the span of the past decade, I was surprised to learn from Robert Wilkins, the producer/director, that this was their first-ever series on Antarctica. Wildlife and travel films, news magazines, and popular science articles are branding the Adélie penguin as the iconic symbol of climate change in Antarctica. Essentially, and rightly so, the Adélie is becoming the "polar bear of the Southern Hemisphere."

Walking across Torgersen Island, the film crew and I passed over what are now hauntingly called *extinct rookeries* where generations of Adélies had fought off skuas, hatched eggs, and nursed their young chicks through fledging. Meandering through one of these rookeries, we rounded a small rocky bluff and encountered a bull fur seal some thirty feet away. It slid a few more feet from us, paused, and then twisting its head in a seemingly impossibly awkward maneuver, peered back over its shoulder toward us. Its facial expression read, "Go ahead, just try coming any closer."

Antarctic fur seals, the only seals of the Southern Ocean with visible external ears, have, like the elephant seal, extended their geographic distribution southward along the peninsula in concert with the warming conditions. Hunters forced these seals, with their strikingly thick fur coats, to the brink of extinction during the late eighteenth and early nineteenth centuries. Sealing in the Southern Ocean was initially, and somewhat paradoxically, triggered by James Cook's circumnavigation of Antarctica in 1772–75. Cook was a stickler for details, and his meticulous nautical accounts included notes on the astounding abundance of seal populations. These numbers proved too tempting to ignore. While the first wave of sealing in the Southern Ocean (1784–1818) focused on the Falkland Islands and Cape Horn region, the industry eventually made its way across the Drake Passage to South Georgia and its surrounding islands. During this period, on Livingston Island in the South

Shetlands alone, sealers slaughtered and skinned a stunning 14,000 fur seals in just five weeks.

Over the next two decades, Russian, American, and British sealers slaughtered millions of fur seals and shipped the skins north. Furriers fashioned the skins into coats, capes, stoles, and hats. Elephant seals were similarly slaughtered in huge numbers, but for their oil rather than their fur. Sealers easily shot or clubbed the huge seals over their heads, then processors lanced and boiled flesh to extract the oil and ship it north in wooden casks. The 1959 Antarctic Treaty, the 1972 Convention for the Conservation of Antarctic Seals (CCAS), and the protective legislation in countries such as Great Britain, Australia, Chile and Argentina, which claim regions of Antarctica within their broad biogeographic ranges have reversed hunting trends. The populations have rapidly recovered and now number in the millions once again.

During hikes on Amsler Island in the austral fall, I've often encountered large bull male fur seals similar to the one on Torgersen Island, surrounded by their harems of much smaller females. Threatened, bull males will stand their ground, growling and lunging—an intimidating sight since they can grow up to six feet long and weigh four hundred pounds. As leader of the 2007 Antarctic Climate Change Challenge mission aboard the *Minerva*, I had to repeatedly warn tourists to keep their distance, reminding them not to confuse sleeping fur seals with the gray rocks scattered around the beaches. One day, I watched a tourist step toward a sleeping fur seal without even realizing it. Fortunately, I was able to alert her to the danger. A seal can inflict a nasty bite that runs a high risk of infection. Indeed, one might end up with a serious medical condition known as "seal finger," or in Norwegian, *spekk-finger*, which translates literally to "blubber-finger." The documented history of this affliction dates from the early sealing days and still occurs at animal parks where handlers work with marine mammals. According to medical reports, seal finger causes edema of the bone marrow, debilitating joint inflammation, and cellulitis. The infection, likely of bacterial origin, is not something to take lightly. Historically, doctors treated seal finger by amputation of the affected digit.[4]

Meandering across the rock-strewn landscape of Torgersen Island, Jen Blum, one of the group of Palmer scientists studying birds or "birders" as they are affectionately called by station staff, explained to the *Globe Trekker* film crew who were setting up for a camera shot of the bull fur seal we encountered, that these seals spend most of the winter at sea feeding primarily on krill. Thus, with krill stocks declining along the central and northern peninsula in tandem with the recession of the annual pack ice, some scientists are concerned that populations in the region may already be decreasing as the fur seals' primary nutritional resource becomes increasingly scarce. In addition, recent studies have revealed that reduced krill populations are tightly correlated with extreme variation in fur seal reproductive output, with females in krill-depleted regions producing fewer pups.[5]

Despite their falling numbers to the north, the seals themselves are responsible for damaging another part of the Antarctic food chain near Palmer Station. As they move into areas previously occupied by colonies of Adélie penguins, fur seal populations have crushed and smothered the native vegetation, endangering its very survival, particularly on Litchfield Island, a nature preserve one mile west of Palmer Station. Perhaps one of the most outstanding features of this pristine island reserve is its rich carpet of mosses that are collectively representative of the species assemblages that occur along the Antarctic Peninsula. Unfortunately, the mosses grow at the lower elevations of the island, coincident with the areas where increasing numbers of Antarctic fur seals are appearing each year. British conservationists have gone so far as to recommend to relevant governing organizations that Antarctic fur seals be removed from their protected status in regions where they are endangering the endemic vegetation of Antarctica. This change in their status would allow an appropriate agency to euthanize a certain number of seals in a given period of time. Breeding colonies are also likely to spread to the Antarctic Peninsula, thereby increasing the pressure on governing bodies to control fur seal populations

⌒

*"Shouldn't we let it into the zodiac?!"* our volunteer dive tender from Palmer Station yelled, turning her pleading eyes toward me. The Adélie had surfaced forty feet from our zodiac, squirming within the jaws of a large leopard seal. The seal released the penguin from its grasp, and once again it repeated its desperate swim toward our zodiac. Just prior to reaching our boat, the seal dove and the penguin once again suddenly vanished underwater in a sucking vortex of teeth. This game of cat and mouse continued for at least another fifteen minutes, clearly a display of dominance put on for our behalf. But one thing that those of us who had experienced scuba diving with leopard seals knew, and that our volunteer dive tender did not, was that if we had allowed the Adélie to climb into our zodiac, the leopard seal would surely have followed suit. A thousand-pound leopard seal would surely swamp the boat. Ultimately, the Adélie's fate was sealed at the moment of its capture: the leopard seal eventually tired of its game, and, tossing the Adélie into the air like a rag doll, tore into its flesh, swallowing it within a minute.

Leopard seals are the most well-known and cautiously respected top predator in Antarctic peninsular waters. They grow to considerable size, with bull males reaching eight feet in length and 1,000 pounds, while females, known as *cows*, are even larger, attaining body lengths of eleven feet and up to 1,300 pounds. More impressive than their sheer size are their massive and muscular necks and heads, the latter equipped with powerful jaws lined with razor-sharp sharklike teeth. Interestingly, their teeth are designed to tear and shred penguin and seal prey. In addition, they are also spaced to facilitate the consumption of krill engulfed with seawater. Leopard seals have a nasty habit of chewing on the rear ends of our inflatable zodiac boats, periodically causing vessels to deflate. Over the years, the Palmer Station boating coordinators have devised heavy plastic cones that fit over the ends of the port and starboard aft rubber compartments of the zodiacs to prevent the seals from penetrating them with their sharp teeth. The sheer force of a seal bite is capable of popping the tough rubber skin of a zodiac as easily as a child popping bubblegum in his or her mouth.

In 2004, a leopard seal popped the two aft sections of a zodiac moored right in front of the station. The back end of the boat sank, immersing the engine in the subtidal and critical krill sampling research gear aboard the zodiac sank into the water. Ryan Wallace, one of the boating coordinators, had to don a cold-water exposure suit and poke around on the seafloor to recover the lost equipment. Then the Palmer Station SkyTrack, essentially a hydraulic crane on wheels, pulled the partially sunken zodiac out of the water so that it could be patched. More recently, one of Ryan's colleagues actually caught a leopard seal on video trying to bite into the side of one of the zodiacs, gnawing at it as a dog would a bone. Fortunately, it couldn't secure enough purchase with its teeth to penetrate the rubber of the highly pressurized compartment, though it did leave behind some nasty-looking tooth scrapings.

My first encounter with a "lep" (as they are referred to reverently in the Antarctic biological vernacular) was typically daunting. It was 1987, and I was working as a postdoctoral research fellow under the direction of Dr. John Pearse. The National Science Foundation had awarded John a grant to study aspects of polar marine invertebrate reproduction, and we were spending several months at McMurdo Station, the site of John's doctoral studies in the mid-1960s. Our research team was scheduled to helicopter to the edge of the pack ice where we would dive and collect echinoderms. It was a glorious Antarctic day, clear and sunny with that stunning blue-tinged brightness that only a polar sun reflecting off ubiquitous snow and ice can generate. Unable to land on or fly over the sea ice due to safety precautions, the pilot of our helicopter gently set us down along the rocky shore. Soon after the engines shut down and the huge rotor blades came to a stop, a group of about twenty emperor penguins meandered over to our helicopter, as if to collectively welcome us to a lovely day on the ice edge.

As our helicopter departed, leaving our team alone on the ice, my dive partner Ron Britton and I carried our scuba and collecting gear over to the ice edge where we donned our dry suits. Ron was an expert diver and experienced field assistant that John Pearse had hired who later became a wildlife biologist for the U.S. Fish and Wildlife Service in Alaska. No sooner had we finished than a large leopard seal burst from the sea, some eight feet in length, with a distinct pattern of spots on the skin below its lower jaw. One must see a leopard seal to fully appreciate it. Unlike the roly-poly fur seal pups or the humorously grotesque elephant seals, leopard seals appear reptilian, almost snakelike, with yellow, tiger-slit eyes and seemingly devilish grins, punctuated with menacing teeth. This seal thrust its chest up against the ice edge in front of us and leered. Our research team huddled to discuss our options. We had invested substantial time and energy to travel to the edge of the pack ice to collect our samples. Therefore, we agreed with little debate to remove our dive suits and re-dress, and to lug all our gear along the ice edge to a new location some distance from where we had encountered the seal.

After trekking for an hour, we reached a new site where the coast seemed clear. Once again, we laboriously donned our dry suits. Ron had seated himself on the edge of the sea ice next to a safety rope we had dropped off the ice edge and anchored to the seafloor. His flippered feet dangled in the sea as he leaned over to pick up his dive mask. Suddenly, without the slightest warning, a large leopard seal, bearing the same pattern of spots we had seen earlier, shot right up between Ron's legs and looked him in the eyes. How I had the presence of mind to grab my camera I will never know, but I actually managed to photograph the seal perched right next to our dive safety line, while Ron's adrenaline was fueling a two-and-a-half back gainer with one hundred and twenty-five pounds of gear strapped to his body. Having recognized the seal as the same one that scared us away from our original collection site, we concluded that it had been swimming right below us as our research team moved along the ice edge, a common behavior seals employ to stalk penguins and other small seals. They usually wait below, hidden until the unsuspecting prey drops into the sea before they attack. Because this seal had surfaced before Ron had descended, we surmised that it was mostly curious rather than interested in a potential meal. However, placing self-preservation above scientific inquiry, we decided not to test this hypothesis and aborted our dive plans for the day

Leopard seals are even more common in the vicinity of Palmer Station than McMurdo Sound. Our research team learned the hard way that our divers should return directly to the zodiac, rather than swim to the nearest shoreline, when a leopard seal is sighted. One day in 2000 when our team was collecting marine invertebrates, the coastal waves were large and forceful, and Katrin Iken, now a marine scientist at the University of Alaska, was diving just off the shore of one of the small rocky islands near the station with Bill Baker. I was piloting the zodiac, and the station boating coordinator was with me to assist as a dive tender when a leopard seal suddenly appeared. The divers in turn immediately made for the shore and scrambled up on to the rocks amidst the rough surf. Katrin, small-boned but tough as nails, was able to ride a wave up on to the shore and perch precariously on top of a rocky platform. Unfortunately, the subsequent incoming wave washed her from said platform, whereupon in full dive gear she plummeted some five feet to the rocks below, landing on her back. Amazingly, she was uninjured, and by this time, I had turned the zodiac over to our skilled boating coordinator who masterfully drove the boat to the shore between swells to retrieve our divers, one by one.

Human encounters with leopard seals are dangerous but not usually fatal. On July 22, 2003, Kirsty Brown, a twenty-eight year old British graduate student, became the first human to be killed by a leopard seal. Kirsty was conducting a routine research dive at Rothera Station on Adelaide Island at the base of the western Antarctic Peninsula. She was snorkeling around dusk, and from below a leopard seal might have mistaken her for a seal or penguin and dragged her down several hundred feet. Leopard seals are rare in this region during the winter and he or she may have been malnourished and more aggressive than normal.

Kirsty's tragic death was a wake-up call for the British Antarctic dive program, which put into place a series of more stringent regulations about diving in the vicinity of leopard seals. Our own dive team now employs a leopard seal recall device, an electronic siren that dive tenders quickly lower into the water if they spot a lep while divers swim below. While no specific studies of their population biology relative to climate change exist, the fact that leopard seals are generalist predators that include Adélie penguins and krill in their diet renders them vulnerable to fluctuations in prey populations along the peninsula. Moreover, they depend to some extent on the annual pack ice in the provision of habitat for hunting and resting between foraging bouts.

Unlike leopard, elephant, and fur seals, Weddell and crabeater seals are highly dependent upon the annual sea ice for hunting and pupping, and thus for their very survival, which makes them, similar to the Adélie penguin, highly vulnerable to the impacts of climate change along the peninsular region of Antarctica. Not surprisingly, with the rapid retreat of the annual sea ice, their numbers along the peninsula are plummeting. Over the twelve years that I conducted my research in the McMurdo Sound region, I often experienced the thrill of having Weddell seals swim with me and my colleagues as we dove under the ice. Named for Sir James Weddell, an early-nineteenth-century commander of British sealing expeditions, Weddells are relatively meek in comparison with leopard seals. When the seals would boldly misappropriate our dive holes as their own, the dive tenders could scare them off so our divers could surface as needed. One year, researchers attached video cameras to the backs of Weddell seals, only to discover, remarkably, that they blow bubbles into the delicate platelet ice to scare out fish, supplementing their diet of krill. Weddells can dive thousands of feet to catch large, deep-water Antarctic cod, as evidenced by the seal that surfaced one day in our dive hut with a huge cod in its jaw. Camping on the sea ice at Granite Harbor during overnight field trips, we would lie in our sleeping bags and fall asleep to the haunting clicks and decrescendo whistles of the Weddells below us.

One of the most fascinating tales of evolutionary adaptation concerns the remarkable ice-chipping teeth of the Weddell seal. These teeth come poignantly into play when pregnant females give birth in the austral spring. Rather than select birthing sites along the leading edge of the fast ice where killer whales and leopard seals routinely hunt, Weddell seals swim up under the sea ice and locate a crack or weakened spot at which they chip away with their teeth. Although they have to make repeated trips to the ice edge to breathe, eventually they create a small hole through which they can breathe, and then, at a more leisurely pace, continue to chip away at the ice hole until it is large enough to pull all eight feet of their thousand-pound bodies up and out of the sea. Here, far removed from predators, they birth and nurture one or two pups. All this is made possible by these remarkable teeth. Unfortunately, this adaptation comes with a price, for the life span of the female Weddell seal is only about half that of other seals (twenty years versus forty). It turns out that the constant wear and tear on their teeth from chipping away at the edges of sea ice to maintain breathing holes eventually renders the seals incapable of feeding, and in the end they starve.

Crabeater seals are about half the size of Weddell seals, with pointed snouts and blond coloration. Similar to the Weddell, one of their most unique features is their dentition. Each tooth has a series of protuberances with spaces between them that render them, with jaws clamped, capable of straining krill from seawater. The name "crabeater seal" is actually a misnomer in that they don't eat crabs. The naming error arose when sealers thought the bits and pieces of crustaceans in the seal's scat came from a diet of crabs, when, in fact, they were the remains of the krill that make up 98 percent of the seal's diet. As the crabeater's dependency on krill is greater than that in the Weddell seal, it is positioned to be even more vulnerable to the krill declines occurring along the peninsula. Nonetheless, the current abundance of this species boggles the imagination. Conservative estimates of upward of fifteen million individuals make this species among the most abundant large mammal on earth.[6] Ironically, as they spend most of their life on the inaccessible offshore pack ice, research into their life history and ecology has proven difficult for marine mammalogists, and they remain among the least studied of the seals. This lack of information is a shame, as their intimate ecological connections to the pack ice and high dependence on krill make them, much like the Adélie penguin, particularly vulnerable to the impacts of climate warming.

❧

*The stomach contents of the Adélie,* teased from its living belly by gently forcing warm water down its gullet, were splayed out in a plastic tray on the laboratory bench top. Jen Blum and her fellow "birder," Kirsty Yeager, were intent on their analysis. I watched them use a dissecting microscope to sort through the muck, finding tiny fish scales or *otoliths,* certain evidence of small fish in the diet. Yet the vast majority of the gut contents were comprised of bits and pieces of partially digested krill. These and other similar observations emphasize the key importance of krill in the Adélie diet, and remind one of the uncomfortable reality that as the krill disappears, so do the Adélies. Later that same day, through the BioLab window, I sighted a humpback whale in the adjacent bay, diving and surfacing, spouting, perhaps feeding. It occurred to me that these two seemingly unconnected events, a pan full of small bits and pieces of krill taken from an Adélie's gut and the joyful play of a baleen whale, were connected. Both animals, one large and one small, were both dependent on a dwindling resource, and both may eventually be driven from these peninsular waters.

Over the years, I have had some incredible encounters with whales along the Antarctic Peninsula. Two large humpback whales once visited my zodiac near Palmer Station. They passed by me so closely that I could almost touch the large barnacles encrusting their heads, and I felt the spray as their flukes hit the sea surface. A different time, on board the cruise ship *Minerva* just off the north-central Antarctic Peninsula, I watched two humpbacks bubble feed for over an hour. Descending repeatedly, they took turns blowing rings of bubbles to concentrate schools of krill, and then they lunged upward through the massed crustaceans. We all watched, mesmerized, as they cleared the surface and strained the seawater for the bounty of krill. The whale biologist on board was particularly overjoyed.

For the most part, the whales that frequent the waters of the Antarctic Peninsula, the sei, right, fin, Minke, and humpback, all collectively belong to the cetacean group known as *baleen* whales. Baleen plates are connected to the whale's upper jaw, and resemble giant combs comprised of parallel rows of flexible plates with frayed edges. Consisting of protein-like keratin, the combs are similar in their composition to human hair and fingernails. These plates are incredibly effective at filtering zooplankton from seawater. Baleen whales engulf seawater then push their tongues up against their upper jaw and force water through the baleen combs, straining out tiny crustaceans such as copepods and amphipods, as well as krill and small fish. They can efficiently process immense quantities of seawater and zooplankton. The blue whale, the largest animal on our planet, is also a baleen whale.

As plankton feeders, the common baleen whales of the Antarctic Peninsula are equally affected by the changing climate as Adélie penguins and other sea ice or krill-dependent species such as Weddell and crabeater seals, and even perhaps the leopard seal. Consequently, marine ecologists from a range of countries are sounding an urgent clarion call to monitor baleen whale populations in this sensitive region. To paraphrase Bill Fraser, we must realize that the top predators of the Antarctic Peninsula—the seabirds, seals, and whales—are the most sensitive indicators of ecosystem change because their life histories allow them to respond to, and to integrate, environmental changes over long periods of time and across broad geographic ranges. But because of the rapidity of climate change, they can't seem to adjust this time. They do indeed act like the proverbial canaries in the coal mine, and the changes in their populations are best interpreted as a wake-up call. While Antarctic ecosystem change may not sound like such a big deal to the average person, it truly is.

# Chapter 8

## Closing the Gap

ANTARCTICA AS A GLOBAL SOLUTION

*D*awn breaks early as the austral fall descends on Palmer Station. I arose before first light, slipped into my pants, shirt, and socks, laced my boots, and donned my jacket, headband, and a pair of gloves before heading out the door. I walked briskly up the well-worn path that wound through the boulder-strewn "backyard" to the base of the glacier. My boots crunched through crusts of fresh frozen snow. Thirty minutes later, breathing hard, I reached the summit of the Marr Glacier. Enveloped in silence, I paused and took in the immense panorama backlit by the dawn. To the west lay islands, endless ocean, and sky. To the south the snow-covered peaks of the Antarctic Peninsula were bathed in alpenglow. Far down the path I had just trod stood Palmer Station, nestled on its tiny peninsula, its inhabitants asleep. At times, the remoteness of Antarctica extracts a pound of flesh.

In February 2010, the ringing of the phone next to my bed seemed, at first, to be part of a dream. But eventually, inevitably, the sound pulled me out of my deep slumber. It was too early in the morning for a phone call. On the other end of the receiver, the tenor of Ferne's voice spoke volumes: "Luke has been in a terrible car wreck! Late last night he was taken by ambulance to the emergency room at Children's Hospital, bleeding heavily from gashes in his left shoulder. He has a cracked pelvis and the bones in his left wrist are shattered." It was horrific to wake in Antarctica to the news that my seventeen-year-old son had skidded off a winding mountain road, collided with a large tree, and then rolled thirty feet down a steep embankment. Some ten thousand miles away, separated by oceans and continents, I imagined my son pinned upside down in a confusion of air bags, fumbling with one functional hand to free himself of the seat belt so he could climb out a broken window. Nobody could blame the neighbor who, fearful, called the police when Luke appeared, covered in blood, screaming for help. "I didn't call you from the hospital last night because I knew there was nothing you could have done," said Ferne. I am not sure I would have had the strength to resist calling her had our roles been reversed.

Over the next ten days, while Luke slowly began to recover, Ferne, essentially a single parent when I am on my research trips, guided Luke through his drug-induced delirium and what must have seemed like countless surgeries and medical tests. In the weeks that followed, everyone at Palmer Station felt my pain and responded, each in his or her unique way. In the spirit of an extended family, the likes of which Antarctic stations are famous for, my colleagues reached out, checked up on me, offered words of comfort, and shared stories of their loved ones who had recovered from car accidents. Joanne Feldman, our station MD, known as "Dr. Jo" around the station, comforted me by sharing her professional insights; like a medical detective on a CSI television episode, she informed me that from what she knew about the physics of broken hips, Luke's pelvis couldn't be too seriously damaged if he had walked to the neighboring house right after his accident. Rebecca Shoop, our station leader, maternal and compassionate, was a constant and gentle source of comfort for me day after day.

A couple of days after Luke's accident, a ten-by-three-foot roll of white poster paper was unfurled by some of the station staff in the galley and laid across a table. Kate Schoenrock, an artistic graduate student, drew the large colorful words "Get Well Luke" on the poster. Penguins, seals, and whales played among the large block letters. Soon everyone at the station had signed their names to Luke's poster, and after dinner all thirty-two of us assembled outside for a photograph with Arthur Harbor and the Marr Glacier as a backdrop. Those in the front row proudly held the poster. In the blink of an eye, the digital image traversed cyberspace to Roberta Challener back in Alabama. The next day, Roberta had a photography shop print the image as a large glossy three-by-four-foot colored poster, and she promptly delivered it to Luke's hospital bedside. For me, the poster will live forever as a gift from the collective hearts of my extended Antarctic family.

Just as that phone call forced me to confront my son's mortality in the face of a devastating accident, my personal observations in Antarctica have served as a wake-up call to a rapidly changing climate. By the end of the century the annual sea ice along the central and northern regions of the western Antarctic Peninsula will have vanished. Adélie penguins will have vanished with it; krill will have been replaced by salps; seafloor organisms will be threatened by rising temperatures, ocean acidification, and invading king crabs; the defensive chemicals that kill pathogens and ward off predators will be lost. The possibility of utilizing chemical compounds found in organisms, millions of years of molecular evolution in the making, would disappear. Certain chemicals, if properly studied, developed, and produced, could provide cures to cancer, fight deadly flu viruses, or reduce an unnerving list of antibiotic-resistant bacteria.

Such discoveries are not flights of fancy. In the chemical ecology program that Bill Baker, Chuck Amsler, and I direct, we have already found chemicals in organisms from the Antarctic seafloor with biomedical potential. For example, the outer body tissues of a bright orange softball-sized Antarctic sea squirt have revealed a novel bioactive chemical we have named *palmerolide* to honor Palmer Station. The U.S. National Cancer Institute (NCI) tested palmerolide against a suite of human cancers ranging from leukemia to lung, colon, ovarian, skin, and breast cancer. When the NCI tested palmerolide against melanoma, the deadliest form of skin cancer, they struck pay dirt—the tiniest quantities of palmerolide killed melanomic skin cancer cells. This represents an advantage in cancer therapy, as healthy noncancerous cells are less likely to be damaged with the low-dosage treatments. A pharmaceutical firm is continuing to study how palmerolide works and to evaluate its prospects for development into a cancer drug. It would be quite ironic to discover a cure for skin cancer in an Antarctic sea squirt that lives directly beneath the hole in the ozone, a massive gap in the atmosphere's ultraviolet (UV) radiation protective layer that seasonally exposes Antarctic tourists, scientists, and station personnel, as well as the entire population of New Zealand, to enhanced levels of ultraviolet radiation.

Bill Baker's drug discovery program at the University of South Florida has also found a protein in an Antarctic red alga that is active against strains of the influenza (flu) virus. The protein prevents infection by interfering with the ability of viruses to attach to cells. In many respects, the marine communities of Antarctica are a natural laboratory for the evolution of chemical diversity. With the exception of coastal icebergs scouring the seafloor, these communities have evolved in a stable environment with constant seawater temperatures and unwavering levels of salinity and oxygen. Isolated for millions of years, the structure of these ancient Antarctic seafloor communities is governed in large part by biological factors such as predation and competition. Scripps Institute of Oceanography's Paul Dayton, the pioneer of Antarctic seafloor ecology, classifies these communities as "biologically accommodated." Under this regime, marine algae, soft corals, sponges, treelike cnidarian hydroids, encrusting colonies of bryozoans, sea squirts, and other immobile marine invertebrates have evolved portfolios of bioactive chemicals to ward off predators and discourage neighboring organisms from encroaching on their territory. This chemical warfare on the seafloor and its resulting biomedical benefits are but humankind's for the taking. To squander such chemical biodiversity is reason alone to consider strategies to mitigate the impacts of rapid climate change on Antarctic marine ecosystems.

Climate change is real. I have seen it with my eyes. Over the past decade, I've lectured broadly to public audiences on the ecological impacts of climate change, and I have witnessed a change in the public perception of this phenomenon. Ten years ago, my speaking invitations came primarily from environmental organizations and teachers and professors of high school and college science classes. At large public speaking engagements, I would share the podium with renowned scientists including Jared Diamond, author of *Collapse: How Societies Choose to Fail or Succeed,* and Mike Tidwell, author of *Bayou Farewell: The Rich Life and Tragic Death of Louisiana's Cajun Coast.* But over the past five years, interest in my polar climate change lectures has expanded to business and civic organizations, including Civitan and Rotary clubs. In Alabama, I have spoken to two hundred members of the Birmingham Downtown Rotary Club, the second-largest chapter in America, and a stronghold of Alabama's top CEOs and business leaders. Word spread, and I subsequently spoke at Birmingham's Shades Mountain Rotary Club and the Tuscaloosa Rotary Club. While I initially fielded audience questions about whether the earth was indeed warming, lately I've found that most are convinced that the earth has demonstratively warmed over the past three decades. Among the populace, a growing percentage believes that while average global air temperatures have increased about 1 to 2 degrees Fahrenheit, certain regions of the earth, including the Arctic and the Antarctic Peninsula, have warmed substantially more than the global average.[1] I now answer questions that center on whether humans are responsible for climate warming or whether it's part of the earth's natural cycle.

I was raised on a diet of objectivity, and so I approach issues with the mindset that I must be convinced by well-grounded data that originates from rigorous peer-reviewed sources. Ironically, what ultimately convinced me that humankind was contributing to rapid global climate change lay in observations of the Antarctic ice sheet. This massive ice sheet, one of only two in the world (the other is in Greenland), coats 98 percent of the Antarctic continent and contains 61 percent of all the freshwater on earth. If the Antarctic ice sheet were to melt, global sea level would rise an estimated two hundred feet.[2] Fortunately, the threat of this major ice sheet melting in its entirety is not an imminent threat to civilization.

The Antarctic ice sheet began to form about forty-five million years ago, the result of a slow, steady, accumulation of snowfall compacted into glacial ice that now extends to a depth of up to three miles. In the 1990s, a team of paleoclimatologists studying how the earth's climate has changed over geological time drilled 10,859 feet into the Antarctic ice sheet near Vostok, a Russian station located in the central region of the east Antarctic ice sheet. Over the course of several months, the scientists' drill bit passed through an impressive 420,000 years of our earth's history. After the scientists had removed the five-inch-diameter ice core and sliced it into manageable three-foot sections, technicians measured the levels of oxygen and carbon dioxide trapped in the ice's tiny air bubbles, which bore a historical record of the composition of the earth's atmosphere. By comparing the ratio of "heavy oxygen" (eight protons and ten neutrons) to "light oxygen" (eight protons and eight neutrons) to a universal standard, the paleoclimatologists were able to reconstruct the history of Antarctic air temperatures. Instances of less-heavy oxygen indicated that air temperatures were cooler.

Led by Jean-Robert Petit, director of the Research in Glaciology and Geophysics of the Environment program at Joseph Fournier University in Grenoble, France, the researchers plotted air temperatures and levels of atmospheric carbon dioxide over the 420,000-year period. Their report[3] revealed that air temperatures and atmospheric carbon dioxide levels rolled in waves across time when graphed, each wave's rise and fall spanning a period of about one hundred thousand years. Climate scientists attribute these waves to changes in the shape of the earth's elliptical orbit around the sun, with the highest temperatures and greatest levels of carbon dioxide occurring during periods when the earth's orbit is less elliptical, and the lowest temperatures and lowest levels of carbon dioxide occurring when the orbit is more elliptical.

But what really caught my attention were the peak levels of carbon dioxide that distinguished each of these waves. At no time in the past 420,000 years had any of the peaks exceeded carbon dioxide values of about 280 to 300 parts per million. However, since the onset of the Industrial Revolution, levels of atmospheric carbon dioxide have climbed well beyond these peaks, attaining 360 parts per million at the time of the 1999 *Nature* publication. Twelve years later, the scientists at the U.S. National Oceanographic and Atmospheric Administration (NOAA) report that as of June 2011, global levels of atmospheric carbon dioxide have attained a level of 393 parts per million—a 24 percent increase above and beyond peak carbon dioxide levels that span almost half a million years. More telling, in 2004, Jean-Robert's team cored even deeper into the Antarctic ice sheet to a depth that coincided with a time period 720,000 years ago. Despite the addition of another three hundred thousand years of atmospheric history, peak levels of past atmospheric carbon dioxide still did not approach current levels. I was convinced: increasing levels of atmospheric carbon dioxide corresponding with the onset and growth of the Industrial Revolution were no coincidence. Fossil fuel emissions—the primary agents behind rising atmospheric levels of carbon dioxide—coupled with unprecedented rates of global deforestation that have reduced a primary reservoir for carbon dioxide, have rapidly increased the level of atmospheric carbon dioxide, a major greenhouse gas. By the end of this century, climate scientists now expect atmospheric carbon dioxide to attain levels as high as six hundred to eight hundred parts per million.

If the evidence for climate change is so overwhelming, why is there so much uncertainty? As a scientist, I believe one contributing factor is that reporters in the media are compelled, and rightly so, to present both sides of an issue. Interviewers on TV or radio seek out "experts" who represent opposing viewpoints about climate change. Putting these experts from each side in front of the camera or behind a microphone places them, figuratively, on equal footing. What is hidden from view is that the scientist discussing how the Industrial Revolution has resulted in a warming planet is effectively speaking on behalf of all the major professional scientific organizations in the world and the vast majority of the earth's scientists. I believe the other represents a tiny camp of scientists or pseudoscientists who have yet to be convinced by mountains of climate change data or simply have another agenda. As recent extreme climate events, including draughts, floods, and unseasonable high and low temperatures, continue to corroborate predictions made in the 2007 report of the Intergovernmental Panel on Climate Change (IPCC), the media is finding it more difficult to interview qualified scientists who deny climate change.

While no single weather event in and of itself provides evidence of global climate change, patterns of such events becoming increasingly severe can speak volumes. Bill McKibben, an environmentalist, writer, and distinguished scholar at Middlebury College in Vermont, and a member of the American Academy of Arts and Sciences,[4] eloquently professes in a tongue-in-cheek way that humankind should continue to bury its collective head in the sand on climate change—to avoid making connections between the record-breaking outbreak of tornados in April 2011 in Alabama, the record-breaking fires in Texas and Arizona in May 2011 that burned more of America than in any other recorded year, the record-breaking snowfall and rain events that generated record-breaking floods along the Mississippi River in April and May 2011, and the record-breaking 2011 droughts in Texas, Oklahoma, and New Mexico. Nor should humankind connect these events to recent record-breaking floods in Australia, Pakistan, and New Zealand, record-breaking melting of Arctic sea ice, a one-hundred-year record-breaking drought in the Amazon, failed grain harvests caused by record-breaking heat in Russia, and the unprecedented loss of blue spruce forests in the northwest United States and Canada brought about through drought and beetle infestations. "Don't connect the dots," McKibben's prose winks.[5]

Furthermore, uncertainty about climate change is propagated because of the way in which humans process information in a modern digital age. People all around the world have become accustomed to acquiring their information rapidly, and instantaneously, in bite-sized chunks. The media has responded accordingly and the consequences have compromised both the scope and sources of scientific information. Moreover, with the explosion of shows on radio, television, and the Internet, people gravitate toward media that provide information compatible with an individual's particular global and political view. Yet an unfortunate outcome of this personalized approach to information acquisition is that many individuals no longer synthesize it from a balance of independent sources. Further confounding an unbiased evaluation of climate change, the same marketing firms that were hired by cigarette companies to plant doubt about the dangers of cigarette smoke have now been hired by the energy industry to do the same with climate change. And climate scientists themselves need to assume some of the blame for the perceived uncertainty about climate change and its impacts—they have not done a good enough job of making their findings and predictions understandable and engaging to the public. More efforts are needed to encourage climate scientists to develop public speaking skills, provide popular lectures, work with the media, and contribute magazine articles and books to the popular literature.

The dramatic environmental shifts now underway along the western Antarctic Peninsula are in many respects the bellwether of global climate change. And yet Antarctica has another important tale to tell. Just as the depths of Antarctica's ice sheet provide evidence of humankind's hand in a rapidly warming global climate, the skies above it provide a story of inspiration, healing, and hope. In 1956, Joseph Farman, an atmospheric scientist working for the British Antarctic Survey, began collecting data at Halley Station, a British Antarctic base on the coast of the Weddell Sea, on the stratospheric ozone (the second layer of the atmosphere, from about six to thirty miles above 'the earth, that protects life on earth by absorbing ultraviolet radiation from the sun). For twenty-six years, the ozone levels recorded by Farman's technicians were rather mundane. Then in 1982, seemingly out of the blue, the ozone level plummeted 40 percent. As a scientist, Farman's response to this dramatic anomaly in his data was appropriate: he simply discounted the reading, attributing the precipitous drop in ozone to a malfunction in the spectrophotometer the scientists used to make the ozone measurements. After all, scientists had established that spectrophotometric measurements are notoriously difficult to conduct at freezing temperatures, and this one was getting old. The operational manual for the Dobson spectrophotometer used for these measurements suggests wrapping the instrument in blankets when taken outside in freezing temperatures.[6] Furthermore, the United States' NASA program had several satellites measuring stratospheric ozone concentrations all around the globe.[7]

To correct the surely spurious reading, Farman simply ordered a new spectrophotometer. But if Farman had expected the following year's ozone measurement to normalize, he was mistaken; once again, the ozone readings were remarkably low. Puzzled, Farman pored over his ozone data from previous years and realized that a decline in atmospheric ozone over Halley Station had actually begun six years earlier, in 1977. The problem now was that Farman had ozone measurements from only a single geographic location. Could it be that stratospheric ozone depletion was localized strictly to Halley Bay? To address this important question the next year, Farman sent his technicians one thousand miles northwest to Faraday, a small British station just to the south of the Palmer Station. The Faraday measurements also revealed a substantial decline in ozone. Armed with evidence of ozone depletion at Antarctic locations a thousand miles apart, Farman reported the discovery of the depletion of atmospheric ozone over Antarctica in the May 16, 1985, issue of *Nature*.[8] Not surprisingly, NASA scientists were chagrined to learn they had missed the most important environmental discovery of the twentieth century. And to make matters worse, the NASA scientists had been sitting on detailed satellite-generated data that showed the ozone depletion. They had missed the discovery because, surprisingly, their software, processing massive amounts of data, was designed to ignore all ozone readings that were considered outside the realm of possibility. The unusually low ozone readings taken by Farman and his colleagues had fallen into NASA's "ignore" category, and they had attributed the data to instrument failure. Once the NASA team realized their mistake, they reanalyzed their data, resulting in the mind-boggling discovery that a hole the size of the United States in the earth's UV-protective ozone was situated over Antarctica.

The consequences of this massive "ozone hole" (it is not actually a hole but a thinning of the ozone layer) were of immediate concern to everyone. First and foremost, biologists worried about the potential impacts of increased ultraviolet radiation on life—especially increased exposure to shorter wavelength UVB (ultraviolet radiation in the 290–320 nanometer wavelength range) radiation known to cause outright damage to DNA. Marine planktonic organisms that thrive in the uppermost layer of the seas surrounding Antarctica were most vulnerable, and because the Antarctic ozone hole seasonally reduced ozone levels in New Zealand, the potential impact of increased UV radiation on humans was also of concern. Andrew Davidson and Lee Belbin, biologists with the Australian Antarctic Division, conducted a series of key experiments on planktonic marine microorganisms at Davis Station. In a series of experiments conducted in outdoor seawater tanks, the two scientists examined the impacts of levels of UV radiation expected under both natural and ozone-deficient conditions on communities of Antarctic marine microbes. Their study revealed that following UV exposure, some species of phytoplankton died while others were unaffected or even flourished. Nonetheless, the overall quantities of phytoplankton in the natural microbial communities exposed to ozone-hole levels of UV radiation declined. Marine bacterial populations were UV tolerant and their populations increased, especially under conditions where UV exposure killed phytoplankton. Davidson and Belbin's study demonstrated that ozone depletion can alter the types and population abundances of organisms that comprise Antarctic marine microbial food webs.[9] Moreover, because fewer phytoplankton survived under ozone-hole UV conditions, Davidson and Belbin surmised that Antarctica's seas may be less capable of storing atmospheric carbon dioxide, which would exacerbate global climate change.

Coastal Antarctic marine invertebrates such as sponges, soft corals, starfish, and sea urchins living on the seafloor fared better under the hole in the ozone. These organisms generally occur deep enough to protect them from seasonal exposure to dangerous UV wavelengths. However, in a study I published with biologist Deneb Karentz of the University of San Francisco in *Antarctic Science*, we report that Antarctic seafloor invertebrates harbor small chemical compounds (mycosporine-like amino acids) that serve as "sunscreens" that protect them from damaging levels of ultraviolet radiation.[10]

While on sabbatical with my family at the University of Otago in Dunedin, New Zealand, in 2006, I arose early each morning to enjoy a cup of coffee and read the *Otago Daily News*. Before we left the United States, I had emailed a colleague who lived in Dunedin to ask him about the austral spring weather. "Sweater weather" was his response, which turned out to be quite the understatement. The proximity of the South Island of New Zealand to Antarctica heavily influences its weather. Constantly chilled and hoping for a warming trend, my family and I would read the *Daily News*'s weather reports daily. Under such close scrutiny, we noticed that the newspaper's weather section included a recommended sun protection factor (SPF) for the day. New Zealanders are acutely aware of the sun's damaging rays and their proximity to the hole in the ozone: the incidence of melanoma is four times higher in New Zealand than in the United States, Canada, or the United Kingdom. While the ozone hole over Antarctica does not actually extend over New Zealand, when the hole disintegrates in the summer as the atmosphere warms and stratospheric clouds are no longer available to facilitate the chemical reactions that break down ozone, ozone-depleted air drifts north of Antarctica, lowering the UV-filtering properties of New Zealand's atmosphere by about 10 percent.

Climate scientists suspect that both global greenhouse gases and the hole in the ozone have influenced aspects of the climate in Antarctica. Since the discovery of the hole in the ozone in 1985, scientists have detected intensified westerly winds around Antarctica, especially during the austral fall. Stronger winds can block air masses from crossing into the interior of the Antarctic continent, and they can have opposing influences on Antarctic sea ice, contributing along with greenhouse-gas warming to the loss of sea ice along the Antarctic Peninsula, while maintaining or slightly increasing sea ice in the Ross Sea. Stronger westerly winds may also be combining with greenhouse gases to raise air temperatures along the Antarctic Peninsula, accelerating the recession of glaciers and the loss of ice shelves.[11] In an alarming study, scientists from the University of Colorado, the National Oceanographic and Atmospheric Administration (NOAA), and NASA predict that should the ozone over Antarctica recover, it could actually amplify warming in the Antarctic interior.[12] They predict this warming because a return to normal levels of ozone would result in sufficient UV absorption of the sun's energy to raise the temperature of the earth's stratosphere up to 16 degrees Fahrenheit. An increase in temperature this severe would break down the large-scale atmospheric circulation pattern that is now shielding Antarctica's interior from warmer air masses to the north.

In April 2008, I hosted a Distinguished Visiting Scholar Lecture at the University of Alabama at Birmingham (UAB) campus for Susan Solomon, an atmospheric chemist from NOAA, in Boulder, Colorado. I had long admired Susan's career and had hoped to meet her in person. Her popular book, *The Coldest March*, first brought her to my attention. In this captivating book, she postulates that unseasonal frigid weather may have played a role in the untimely deaths of British explorer Sir Robert Scott and his four colleagues Edward Wilson, Henry Bowers, Lawrence Oates, and Edgar Evans, on the return leg of their epic march to the South Pole from November 1, 1911, to March 29, 1912. I was equally impressed with Susan's professional accolades, which include the National Medal of Science (the highest award for a scientist in the United States), membership in both the U.S. and European National Academy of Sciences, and in her capacity as a member of the Intergovernmental Panel on Climate Change, she shared the 2007 Nobel Peace Prize with former Vice President Al Gore.

Even more impressive is Susan Solomon's love for and dedication to science. She grew up watching *The Undersea World of Jacques Cousteau* and actively mentors early-career women in the sciences. In 2008, *Time* named Susan Solomon one of the top hundred most influential people in the world. But her greatest achievement was her discovery of the cause of the ozone hole over Antarctica. Susan had theorized that human-made chlorine compounds—discovered in 1974 by Sherwood Rowland, Mario Molina, and Paul Crutzen to have a role in ozone depletion—caused ozone-destroying chemical reactions on the surfaces of stratospheric ice clouds that form in the unique low winter temperatures above Antarctica (a frigid −88 degrees Fahrenheit or lower). The National Science Foundation selected Susan to lead the 1986–87 U.S. National Ozone Expedition to McMurdo Station, where I was conducting my postdoctoral research at the time. When Susan's team discovered that levels of chlorine oxides (chemical compounds which are produced when chlorine reacts with ozone) in the atmosphere over Antarctica were one hundred times higher than expected, she had the evidence she needed to prove her theory correct. Furthermore, chlorine was a well-known component of chlorofluorohydrocarbons (CFCs), a group of common compounds used as refrigerants, propellants in aerosol sprays, dry-cleaning agents, and cleaners of electronic components. (Freon is an example of a CFC.) Humankind was thus responsible for the ozone hole. Despite the earth's immense size, the perception of its resilience was no longer sacrosanct.

At the same time that Susan's team was solving the riddle of the hole in the ozone, governments were laying the groundwork for an inspirational and unprecedented development in international diplomacy. In September 1987, representatives of twenty-four nations gathered at the headquarters of the International Civil Aviation Organization in Montreal, Canada and finalized the details of a global ozone treaty. The official treaty, negotiated under the umbrella of the United Nations Environmental Program, was entitled *The Montreal Protocol on Substances that Deplete the Ozone Layer*, and parties signed and ratified it on September 16, 1987. Among the signees representing their countries were Margaret Thatcher and Ronald Reagan. Abbreviated the Montreal Protocol, the treaty recognized that worldwide emissions of certain substances—primarily CFCs, but also halons, carbon tetrachloride, and methyl chloroform—can deplete and otherwise modify the ozone layer in a manner that causes adverse effects to human health and the environment. The treaty also stipulates that nations are to phase out the production and consumption of these compounds by the year 2000 (methyl chlorine by 2005). President Ronald Reagan wrote in his statement on signing the treaty: "The protocol marks an important milestone for the future quality of the global environment and for the health and well-being of all peoples of the world. . . . It is a product of the recognition and international consensus that ozone depletion is a global problem, both in terms of its causes and effects."[13] Over the next decade, subsequent meetings held in London (1990), Copenhagen (1992), Vienna (1995), Montreal (1997), and Beijing (1999) facilitated a variety of amendments to the original Montreal Protocol, including backing up the global phase-out of CFCs from 2000 to 1995 because of growing concerns about increasing ozone depletion. Meetings have continued at an aggressive pace—the twenty-third meeting of the "Parties to the Montreal Protocol" took place in November 2011 in Bali, Indonesia.

The Montreal Protocol forged new ground in international agreements and cooperation by becoming one of the first treaties to include incentives for nations to become treaty signatories while also endorsing trade sanctions as a mechanism to leverage participation. Representatives from the negotiating nations justified the case for trade sanctions against offending nations on the rationale that the depletion of the earth's ozone layer is best addressed at the global level, and that all governments must be incentivized to participate. Without trade sanctions, substantial financial incentives existed for nonsignatory nations to mass produce CFCs to meet a "demand gap" caused by the cessation of the production of CFCs and other banned substances by participating nations. It all worked. Participating countries passed legislation preventing the production of ozone-depleting substances. By 1996, the world's industrial countries had banned the use of CFCs (a small number of CFCs deemed medically essential were exempted), and treaty parties targeted 2010 as the year that CFCs would be banned worldwide. Additionally, national and international financial investments have facilitated the successful development and production of alternative chemical compounds that are not destructive to ozone. And with the assistance of a Montreal Protocol Fund established by developed countries, almost $2 billion has been invested to date to help underdeveloped countries instigate technologies to transition away from ozone-depleting chemical compounds. And best of all, twenty-four years after its initial ratification, the number of treaty signatories has increased almost eightfold to over 180 countries. Given such a vast buy-in, the Montreal Protocol has in every sense become a model of global agreement. Kofi Annan, the secretary general of the United Nations from 1997 to 2006 and corecipient of the 2001 Nobel Peace Prize, captured this sentiment in his proclamation: "Perhaps the single most successful international agreement to date has been the Montreal Protocol."

༄

*Hundreds of college students,* faculty, and community members packed the Jemison Concert Hall in the UAB Alys Robinson Stephen's Performing Arts Center to hear Susan Solomon's April 8, 2008, lecture "A World of Change: Climate Yesterday, Today, and Tomorrow." During her lecture, she outlined in layperson's terms how climate change works and how it has been difficult for individuals and nations to deal with—a phenomenon she refers to as "climate gridlock." She also addressed common misconceptions about climate change such as the confusion between weather, which is the conditions of the atmosphere over the short-term, and climate, which is how the atmosphere behaves over the long term. Based on her rapport with her audience, I could tell that Susan, like me, realized that her role as a climate-change scientist was to inform people about what the problem is, not to steer people's decisions. I have seen audiences quickly discount scientists who present themselves as dogmatic, opinionated, or political. Immediately following Susan's lecture, she joined a panel of experts on the stage who had been invited to discuss climate change and answer audience questions. Among the panel experts was Henry Pollack, a professor of geology at the University of Michigan and author of *A World Without Ice.* Henry has traveled the world, drilling bore holes in the earth's crust and using highly sensitive thermometers to make detailed measurements of the planet's subsurface temperatures. The bore holes extend to a depth of 1,500 feet and because signals from the earth's surface temperatures travel below the earth's crust and are preserved in soil and rock, the readings provide an archive of how the earth's temperatures have changed over the last thousand years. Henry explained to the audience that he and colleagues Shaopeng Huang and Po-Yu Shen of the University of Western Ontario had measured subsurface temperatures at 616 locations around the globe, and their findings

indicated that while average temperatures in the Northern ⌐
have risen 2 degrees Fahrenheit over the last five hundrec
vast majority of this increase—slightly over half of the 2 degrees—
occurred entirely in the twentieth century.[14] What they've found rein-
forces the same story scientists revealed in cores bored into the Antarctic
ice sheet and through ancient coral reefs—the earth's temperatures have
risen unusually quickly over the past century.

I drove Susan to the airport, and on the way we chatted about the
future of the ozone hole over Antarctica. She told me that the international regulation of CFCs and other ozone-destroying compounds had
exceeded even her expectations, and that impressive reductions in their
global production and use have been made. As such, she said, over recent years she and other atmospheric chemists around the world had
seen a dramatic decrease in the rate of ozone depletion over Antarctica.
In fact, Susan said enthusiastically, "Scientists following levels of ozone
depletion are now observing that the growth of the hole is in remission,
with the depletion not getting any worse." Susan was on the right track.
Just two years later, three hundred scientists representing the United
Nations Environmental Program (UNEP) published a peer-reviewed report entitled *Scientific Assessment of Ozone Depletion 2010*.[15] In their
report, the scientists articulate that the 1987 Montreal Protocol that initiated the phase-out of CFCs and other ozone-destroying substances had
been successful, and that levels of global ozone, including the ozone in
the stratosphere over Antarctica, were no longer decreasing. And most
exciting of all, the UNEP scientists now expect that global ozone will be
restored to 1980 levels between the years 2045 and 2060, even slightly
earlier than the UNEP had expected.

Achim Steiner, executive director of the United Nations
Environmental Program and under-secretary-general of the United
Nations, poignantly emphasized that had there never been a Montreal

Protocol, levels of ozone-depleting chemicals would currently be ten times greater in the atmosphere, which could have caused several million more cases of cancer, over one hundred million more eye cataracts, and broadly compromised the human immune system and instigated planet-wide damage to agriculture and wildlife. The story of the discovery, mitigation, and pending recovery of the hole in the ozone over Antarctica offers a vision of hope for the future of our planet. Despite the complex challenges of addressing the global regulation of carbon dioxide and other greenhouse gases, it is remarkable that a coalition of over 180 developed and underdeveloped countries could come together to heal a globally induced problem that altered the fundamental nature of the planet's atmosphere.

In 2001, I first tied the bowline of my zodiac to the steel mooring pin set firmly in a rock along the shore at the site of the old Palmer Station, the predecessor of the current location bearing that name, on the southwestern coast of what is now Amsler Island. Only about a mile from Palmer Station, science support staff built and occupied Old Palmer as a temporary station between 1965 and 1968, essentially serving as a staging area for the building of the new station. It was maintained for many years as an emergency backup station should Palmer Station need to be abandoned. But Old Palmer is no more. Where a small group of temporary buildings were once clustered, only a rock-strewn landscape remains. The demolition and clean-up effort of Old Palmer—conducted in the early 1990s as part of an environmental effort by the United States to reduce the footprint of its Antarctic facilities—was so complete that it is now impossible to find the slightest trace of human habitation.

Now, a decade after my first visit to the abandoned station site, I clamber up and over the boulders, dodge the dive-bombing skuas, and take a circuitous route around several fur seals. A short climb takes me to the summit of Amsler Island. Below me, to the south, a small island called Elephant Rocks bustles with a breeding colony of barking elephant seals, the product of a recent range-extension from the warmer subantarctic. Torgerson Island, which is also within earshot, is oddly quiet, its former Adélie colonies in large part permanently abandoned. To the west, the tongue of the Marr Glacier that had so recently extended seaward, covering Amsler Island, has dissipated, its final vestiges melting rapidly to reveal rock and dirt. As I descend from the summit, I pause to gaze into the distance at the cluster of deep-blue, tan-roofed buildings that make up Palmer Station. What sort of a world, I wonder, will future generations of Antarctic scientists find when they come to this remarkable place? And when they gaze over this landscape, will they be reminded how this place, this peninsula, these ecosystems, served as a wake-up call to jump-start the technological, societal, and political paths to a sustainable planet?

# Notes

## Chapter 1

1. The website for Mustang Survival, mustangsurvival.com.

## Chapter 2

1. Sarah Gille, "Warming of the Southern Ocean since the 1950s," *Science* 295 (2002): 1275–77.
2. John Fyfe and Oleg Saekno, "Human-induced change in the Antarctic Circumpolar Current," *Journal of Climate* 18 (2005): 3068–73.
3. Lisa Beal, et al., "On the role of the Agulhas system in ocean circulation and climate," *Nature* 472 (2011): 429–36.
4. Peter Convey, et al. "Antarctic climate change and the environment," *Antarctic Science* 21 (2009): 541–63.
5. Michael Meredith and John King, "Rapid climate change in the ocean west of the Antarctic Peninsula during the second half of the 20th century," *Geophysical Research Letters* 32 (2005): 32: L19604.
6. David Vaughan, et al., "Recent rapid regional climate warming on the Antarctic Peninsula," *Climate Change* 60 (2003): 243–74.
7. Eric Steig, et al., "Warming of the Antarctic ice-sheet surface since the 1957 International Geophysical Year," *Nature* 457 (2009): 459–62.
8. Ibid.
9. Peter Convey, et al. "Antarctic climate change and the environment."
10. David Vaughan, et al., "Recent rapid regional climate warming on the Antarctic Peninsula."
11. "Runaway ice chunk in Antarctica worries scientists," *Associated Press/The New York Times* (March 26, 2008): http://www.nytimes.com/2008/03/26/health/26iht-ice.1.11432642.html.
12. Stanley Jacobs, et al., "Observations beneath Pine Island Glacier in West Antarctica and implications for its retreat," *Nature Geoscience* 3 (2010): 468–72; Andrew Shepherd, et al., "Inland thinning of Pine Island Glacier, West Antarctica," *Science* 291 (2001): 862–64.
13. Ibid.

14. David Vaughan, et al., "Recent rapid regional climate warming on the Antarctic Peninsula."

15. Jonathan Bamber, et al., "Reassessment of the potential seal-level rise from a collapse of the west Antarctic ice sheet," *Science* 324 (2009): 901–03.

16. Hamish Pritchard, et al., "Antarctic ice-sheet loss driven by basal melting of ice shelves," *Nature* 484 (2012): 502–05.

17. Eric Rignot, et al. "Acceleration of the contribution of the Greenland and Antarctic ice sheets to sea level rise." *Geophysical Research Letters* 38 (2011): L05503.

18. BBC News, Science & Environment, "Polar ice loss quickens, raising seas," (March 9, 2011): www.bbc.co.uk/news/science-environment-12687272

19. "Contribution of working groups I, II, and III to the Fourth Assessment of the Intergovernmental Panel on Climate Change," *Climate Change 2007 Synthesis Report.* IPCC 446 (2007).

20. Alison Cook, et al., "Retreating glacial fronts on the Antarctic Peninsula over the past half-century," *Science* 308 (2005): 541–44.

21. Edgeworth David and Raymond Priestley, "British Antarctic Expedition 1907–09 Reports on the Scientific Investigations," *Geology* 1 (1914).

## Chapter 3

1. Thomas Hodgson, "On collecting in the Antarctic seas. National Antarctic Expedition, 1901–04," *Natural History* 3 (1907): 1–10.

2. Wes Yoshida, et al., "Pteroenone: a defensive metabolite of the abducted Antarctic pteropod *Clione antarctica*," *Journal of Organic Chemistry* 60 (1995): 780–82.

3. James McClintock and John Janssen, "Pteropod abduction as a chemical defense in a pelagic Antarctic amphipod," *Nature* 346, (1990): 462–64.

4. Thomas H. Maugh II, "Science/Medicine: Crustacean carries its own bodyguard to ward off predators," *Los Angeles Times,* (August 6, 1990): http://articles.latimes.com /1990-08-06/local/me-168_1_predator-fish.

5. "Ocean Glider completes first-ever ocean crossing," National Oceanic and Atmospheric Administration, (2010): http://oceanservice.noaa.gov/news/weeklynews/dec09 /glider.html.

6. Martin Montes-Hugo, et al., "Recent changes in phytoplankton communities associated with rapid regional climate change along the Western Antarctic Peninsula," *Science* 323 (2009): 1470–73.

7. Wayne Trivelpiece, et al., "Variability in krill biomass links harvesting and climate warming to penguin populations changes in Antarctica," *Proceedings of the National Academy of Sciences* 108 (2011): 7625–28.

8. Oscar Schofield, et al., "How do polar marine ecosystems respond to rapid climate change," *Science* 328 (2010): 1520–23.

## Chapter 4

1. Peter Brueggerman, "Diving under Antarctic ice: a history." *Technical Report* (2003): http://escholarship.org/uc/item/1n37j685.

2. Charles Amsler, et al., "Vertical distributions of Antarctic Peninsular macroalgae: cover, biomass, and species composition," *Phycologia* 34 (1995): 424–30.

3. Michael Meredith and John King, "Rapid climate change in the ocean west of the Antarctic Peninsula during the second half of the 20th century," *Geophysical Research Letters* 32 (2005): L19604.

4. Lloyd Peck, "Extreme seasonality of biological function to temperature in Antarctic marine species," *Functional Ecology* 18 (2004): 625–30.

5. David Barnes and Lloyd Peck, "Vulnerability of Antarctic shelf biodiversity to predicted regional warming," *Climate Research,* 37 (2008): 149–63.

6. Suzanne Gatti, "The role of sponges in high Antarctic carbon and silicon cycling—a modeling approach," Ph.D. Thesis, Bremerhaven University, Bremerhaven, Germany. *Ber Polar Meeeresforsh* (2002): 434.

## Chapter 5

1. Elizabeth Kolbert, "The acid sea," *National Geographic* (April 2011).

2. Victoria Fabry, et al., "Ocean acidification at high latitudes: The bellwether," *Oceanography* 22 (2009): 161–71.

3. Carlos Pelejerto, et al., "Paleo-perspective on ocean acidification," *Trends in Ecology and Evolution* 25 (2010): 332–44.

4. Ken Caldeira and Michael Wickett, "Ocean model predictions of chemistry changes from carbon dioxide emissions on the atmosphere and oceans," *Journal of Geophysical Research* 110 (2005): CO9S04.

5. James Orr, et al., "Anthropogenic ocean acidification over the twenty-first century and its impacts on calcifying organisms," *Nature* 437 (2005): 681–86.

6. Philip Munday, et al., "Ocean Acidification impairs olfactory discrimination and homing ability of a marine fish," *Proceedings of the National Academy of Sciences* 106 (2009): 1848–52.

7. Maoz Fine and Dan Tchernov, "Scleractinian coral species survive and recover from decalcification," *Science* 315 (2007): 1811.

8. Ilsa Kuffner, et al., "Decreased abundance of crustose corraline algae due to ocean acidification," *Nature Geoscience* 1 (2008): 114–17.

9. Thomas Huxley, *A Manual of the Anatomy of Invertebrate Animals,* J & A Churchill: London (1877).

10. Mary Sewell and Gretchen Hoffman, "Antarctic echinoids and climate change: a major impact on brooding forms," *Global Climate Change* 17 (2011): 734–44.

11. So Kawaguchi et al., "Will krill fare well under Southern Ocean acidification?" *Biology Letters* 7 (2010): 288–91.

12. *Climate change 2007: Synthesis report,* "Contribution of working groups I, II, and III to the fourth assessment of the Intergovernmental Panel on Climate Change," IPCC 446 (2007).

## Chapter 6

1. Sarah Purkey and Gregory Johnson, "Warming of global abyssal and deep Southern Ocean waters between the 1990s and 2000s: contributions to global heat and sea level rise," *Journal of Climate* 23 (2010): 6336–51; Sarah Gille, "Decadal-scale trends in the Southern Hemisphere Ocean," *Journal of Climate* 21 (2008): 4749–65.

2. Matthew Lebar, et al., "Cold-water natural products," *Natural Products Reports* 4 (2007): 774–97.

3. Eric Niler, "King Crabs Invade Antarctic Waters," Discovery News (February 8, 2011): http://news.discovery.com/animals/king-crabs-antarctic-waters-110208.html.

4. Craig Smith, et al., "A large population of king crabs in Palmer Deep on the west Antarctic Peninsula shelf and potential invasion impacts," *Proceedings of the Royal Society B* 279 (2012): 1017–26.

## Chapter 7

1. John Turner, et al., "A positive trend in western Antarctic precipitation over the last 50 years reflecting regional and Antarctic wide atmospheric circulation changes," *Annals of Glaciology* 41 (2005): 85–91.

2. James McClintock, et al., "Southerly breeding gentoo penguins for the eastern Antarctic Peninsula: further evidence for unprecedented climate change," *Antarctic Science* 22 (2010): 285–86.

3. Ralph Bovard, "Injuries to avian researchers at Palmer Station, Antarctica from penguins, giant petrels, and skuas," *Wilderness and Environmental Medicine* 11 (2000): 94–98.

4. Tania Hunt, et al., "Health risks for marine mammal workers," *Diseases of Aquatic Organisms* 81 (2008): 81–92.

5. Keith Reid and John Croxall, "Environmental response of upper trophic level predators reveals a system change in an Antarctic marine ecosystem," *Proceedings of the Society of London B* (2001): 268, 377–84.

6. Walter Gilbert and Albert Erickson, "Distribution and abundance of seals in pack ice of the Pacific sector of the Southern Ocean," in G.L. Llano (ed.), *Adaptations Within Antarctic Ecosystems*, Washington DC: Smithsonian Institution (1977): 703–40.

## Chapter 8

1. David Vaughan, et al., "Recent rapid regional climate change on the Antarctic Peninsula," *Climate Change* 60 (2003): 243–74.

2. Oleg Anisimov, et al., "Models of thermal expansion," (Table 11.3), IPCC, TAR, WG1 (2001).

3. Jean-Robert Petit, et al., "Climate and atmospheric history of the past 420,000 years from the Vostok ice core, Antarctica," *Nature* 399 (1999): 429–36.

4. Bill McKibben, "A link between climate change and Joplin tornadoes? Never!" *The Washington Post* (May 23, 2011): http://www.washingtonpost.com/opinions/a-link-beteween-climate-change-and-joplin-tornadoes-never/2011/05/23/AFrVC49G_story.html.

5. Ibid.

6. Robert Evans, "Operational handbook—ozone observations with a Dobson Spectrophotometer," NOAA/ESRL Global Monitoring Division (2006): ftp://ftp.cmdl.noaa.gov/dobson/allin1.pdf.

7. NASA, "Space-based measurements of ozone and air quality in the ultraviolet and visible," http://ozoneaq.gsfc.nasa.gov/.

8. Joseph Farman, et al., "Large losses of total ozone in Antarctica reveal seasonal $ClO_x/NO_x$ interaction," *Nature* 315 (1985): 207–10.

9. Andrew Davidson and Lee Belbin, "Exposure of natural Antarctic marine microbial assemblages to ambient UV radiation: effects on the marine microbial community," *Aquatic Microbial Ecology* 27 (2002): 159–74.

10. James McClintock and Deneb Karentz, "Mycosporine-like amino acids in 38 species of subtidal marine organisms from McMurdo Sound, Antarctica," *Antarctic Science* 9 (1997): 392–98.

11. Peter Convey, et al., "Antarctic climate change and the environment," *Antarctic Science* 21 (2009): 541–63.

12. Judith Perlwitz, et al., "Impact of stratospheric ozone hole recovery on Antarctic climate," *Geophysical Research Letters* 35 (2008): L08714.

13. Ronald Reagan, "Statement on signing the Montreal Protocol on ozone-depleting substances," Ronald Reagan Presidential Library, National Archives and Records Administration (April 5, 1988): http://www.reagan.utexas.edu/archives/speeches/1988/88apr.htm.

14. Shaopeng Huang, et al., "Temperature trends over the past five centuries reconstructed from borehole temperatures," *Nature* 403 (2000): 756–58.

15. WMO (World Meteorological Organization), "Scientific Assessment of Ozone Depletion 2010," *Global Ozone Research and Monitoring Project* 52 (Geneva, Switzerland: 2011).

# Index

Black Island, 30
Blake, Dan, 159
Blum, Jen, 184, 192
borks, 66–8
Bosch, Sid, 57–9
Bowers, Henry, 29, 210
British Antarctic Survey, 20, 52–3, 75, 206
brittle stars, 4, 133–4, 156, 159
Britton, Ron, 187
brooding, 11, 112, 143, 178
Brown, Kirsty, 189, 192
Brown Bluff, 174
bryozoans, 89, 124, 200
bubble feed, 192
Byrd, Adm. Richard E., 18

calcite, 121, 125, 132–4
calving events, 42–3
canary-in-the-coal-mine, 199
Cape Armitage, 100
Cape Evans, 103–4
Cape Horn, 135, 182
Cape Royds, 168
carbon dioxide, 70, 76, 120–1, 126, 128,
    130, 132, 138, 202–4, 208, 216
carbon tetrachloride, 212
Casey Base, 179
*Centre National de la Recherche
    Scientifique*, 4
Challener, Roberta, 150–1, 153–5, 161,
    199
check-out dives, 91, 106
chemical defense, 7, 69, 98, 104–5,
    108–10
chemical ecology, 20, 22–3, 65, 83, 108,
    110, 128, 133, 199
chlorofluorohydrocarbons (CFCs), 211–13,
    215
Christine Island, 169
cilia, 11
*Cinachyra antarctica*, 113
clams, 89, 122, 132, 134, 136, 142, 156,
    159
Clarence Island, 42
climate change, 37, 39, 49, 54, 92, 113,
    137–8, 142
    and bacteria, 111
    and baleen whales, 193

and deep corals, 124
and elephant seals, 181
and hourglass dolphins, 20
human-induced, 20, 113, 124, 201–13
and icebergs, 44, 108
and invasive species, 158
and king crabs, 141
in the northern region, 84
and penguins, 170–1, 174, 176, 182,
    189–90
and plankton, 71–2, 85–6
reality of, 201, 204–8
and snowfall, 170–1
*See also* climate warming
climate gridlock, 214
climate warming, 76, 131, 145, 172, 175,
    177, 179, 191, 201
    and the Adélie penguin, 191
    Antarctic Peninsula as poster child for,
        47, 131–2
    human responsibility for, 201–13
    and icebergs, 45
    and king crabs, 149
    and limpet populations, 10
    and marine invertebrates, 112–13
    and petrels, 177–9
    and seaweeds, 108–10
    and surface seawater temperatures, 107
    and Wilkins Ice Shelf, 53
    *See also* climate change
*Clione antarctica*, 65–6
$CO_2$ problem, the other, 10, 120
Cold Water Neoprene Immersion Suits
    (Gumby Suits), 14–16
Cold-weather clothing, 29, 152
community succession, 44
Convention for the Conservation of the
    Antarctic Seals (CCAS), 183
Convention on the Conservation of
    Antarctic Marine Living Resources
    (CCAMLR), 138
Cook, James, 182
copepods, 70–1, 78–81, 193
corals, 138, 154, 215
    hard, 122–7
    soft, 44, 71, 83–4, 89, 106, 112, 122–3,
        159, 200, 208
Cormorant Island, 169